EARTH SCIENCES IN THE 21ST CENTURY

AF837271

# DEGLACIATION

# PROCESSES, CAUSES AND CONSEQUENCES

# EARTH SCIENCES IN THE 21ST CENTURY

Additional books in this series can be found on Nova's website under the Series tab.

Additional e-books in this series can be found on Nova's website under the eBooks tab.

EARTH SCIENCES IN THE 21ST CENTURY

# DEGLACIATION

## PROCESSES, CAUSES AND CONSEQUENCES

MARCELLA BOONE
EDITOR

nova
science publishers
New York

**Library of Congress Cataloging-in-Publication Data**

ISBN: 978-1-53612-501-6

*Published by Nova Science Publishers, Inc. † New York*

# CONTENTS

# PREFACE

Global climate during the Quaternary has been deeply influenced by glacial-interglacial oscillations. As a result, the Earth has experienced alternations between warm and stable climatic periods – coinciding with interglacials, and cold and highly variable climatic intervals – coinciding with glacials. In a suborbital timescale, climate oscillations were maximal during glacial onsets and, very especially, during deglaciations. Previous deglaciation events were associated with diverse changes in earth's atmospheric, physical and biotic environments and an examination of such interrelationships requires the establishment of a competent temporal framework. Accordingly, chapter one examines dating methods that can be used to provide temporal constraints to continental deglaciation events. To illustrate the range of considerations that could accompany a deglaciation dating study, the chapter also gives a brief outline of a case study conducted in western Canada to constrain the Late Pleistocene retreat of the Laurentide Ice Sheet from the region. Chapter Two reports on the influence of deglaciations in the mid-latitude European climate. Chapter Three aims to highlight the influence of global and regional paleoceanographic changes on the deglaciation of the marine based Barents ice sheet since the last glacial maximum (LGM) until the onset of marine environment in the Holocene.

Chapter 1 – Assignment of a temporal scale to deglaciation events is elemental to the understanding of processes involved during and subsequent to the retreat of continental ice sheets. Dating the recession of ice sheets, however, is complicated by the fact that ice sheets cease to exist following deglaciation, leaving behind no tangible remnants of the ice itself that can directly be analyzed. Hence, dating exercises essentially aim to place constraints on ice sheet recession by examining proxy records that resulted from or are associated with the former existence of ice sheets. A deglaciation timescale enables rates of postglacial geomorphic processes as well as the pace of environmental evolution to be determined. Additionally, such chronologies facilitate the analysis of variations in the distribution of biotic communities over time, in conjunction with changes in climatic conditions and other aspects of the environment. In areas where isostatic rebound occurred following glacial unloading, accurate dating allows geophysical models of viscoelastic responses to be calibrated, including changes in sea level. This chapter provides a review of dating methods that have been used to constrain the retreat of Late Pleistocene ice sheets from formerly glaciated regions following the Last Glacial Maximum, with a focus on North America. Chronological methods explored include cosmogenic-nuclide exposure, luminescence, radiocarbon and varve dating techniques. In each instance, the principles behind the dating technique and the types of material that can be dated are examined. The chapter concludes with a brief outline of a case study conducted in western Canada to constrain the Late Pleistocene retreat of the Laurentide Ice Sheet from the region.

Chapter 2 – Global climate during the Quaternary has been deeply influenced by glacial-interglacial oscillations. Since the onset of glaciations, the Earth has experienced alternations between warm and stable climatic periods – coinciding with interglacials, and cold and highly variable climatic intervals – coinciding with glacials. In a suborbital timescale, climate oscillations were maximal during glacial onsets and, very especially, during deglaciations. It is very obvious that the extension of ice sheets has a direct influence on high-latitude climate, both in continents and oceans; contrary to this, climate in middle latitudes is not so influenced by global ice volume or the total extension of ice sheets, but by episodes of rapid variation of sea ice coverage. Since early Pleistocene the most pronounced stadial (cold) events in the mid-latitude North Atlantic occurred during deglaciations, both during glacial/interglacial transitions – Terminations – and episodes of lesser ice sheets depletion. These severe cooling events were followed by very rapid, short-lived warming episodes that, when coinciding with Terminations of climatic cycles, initiated the climatic optimum of the following interglacial. Such millennial scale oscillations affected not only sea surface temperature, but also continental climate as far as the eastern Mediterranean basin. Most deglaciation onsets coincided with maxima in Northern Hemisphere summer insolation. Enhanced heat budget facilitated rapid ice sheets collapse and allowed the release of huge amounts of fresh water to the ocean, triggering the reduction or even the shutdown of North Atlantic deep water formation, which resulted in the advection of subpolar waters along the western European margin and the associated stadial. After deglaciation, once the freshwater perturbation stopped, the meridional overturning circulation rapidly resumed and warm surface water recirculated again in the North Atlantic, producing the fast temperature rise and initiation of interglacial climatic optimum across Europe.

Chapter 3 – According to the prevailing view, the Barents Sea shelf and surrounding archipelagoes were covered by a grounded ice sheet united with the Scandinavian ice sheet during the Last Glacial Maximum (LGM). Its decay started in the deep marginal troughs, the Bear Island Trough in the southwest, Kvitøya and Franz Victoria troughs in the north, ca. 19-18 cal ka BP with development of ice streams and related iceberg calving over the shelf break. Sediment cores and shallow boreholes recovered two distinct units of glaciomarine sediments. The lower Unit III is represented by a glaciomarine diamicton, a disordered mixture of coarse-grained iceberg rafted debris (from sand to pebbles and boulders), and very fine-grained clay (dominated by <0.001 mm fraction) released from the melting glacier ice. It overlies directly the eroded Mesozoic bedrock surface or non-marine basal moraine deposits (glacial till) in shelf depressions. The diamicton was deposited during the early deglaciation stage, which was apparently diachronous in the study area. The faunal and stable isotope data from the Kvitøya Trough supported by AMS-$^{14}C$ dates suggest the older age of the early deglaciation stage in this narrow eastern Svalbard trough. Similar dates were previously published for the St. Anna Trough. However, the deglaciation of local marine-based ice sheet margins continues e.g., within the Franz Josef Land up to date with recent diamicton deposition. The diamicton is commonly overlain by Unit II with a sharp contact in between, deposited during the late deglaciation stage, when the iceberg calving and rafting ceased owing to glacier margins retreat to the land. The Unit II is mainly composed of fine-grained silty clay with thin fine-grained sand layers and lenses. It is often characterized by thin brown and gray lamination. Both laminated and lenticular sedimentary structures suggest pulsating deposition from glacier meltwater suspension flows likely accelerated by sea ice freezing brines cascading. Coarse IRD is rare or absent. Undulating seismic facies occur in the southern Barents Sea resembling sedimentary waves possibly indicating contourite origin of the thick Unit II. The upper boundary of the Unit II with overlying Holocene marine sediments appears to be also diachronous. Since the LGM until the Recent, the deglaciation was fostered by relatively warm surface and subsurface Atlantic water inflow into the Barents Sea through the western and northern deep troughs, respectively. The glacio-eustatic sea-level rise led to Atlantic water penetration beneath the glacier ice resulting in its melting from below thus promoting ice streams activity during early deglaciation. The distribution of iceberg rafted debris and its petrography depended on the provenances drained by ice streams as well as on the icebergs drift routes and their melting sites. Severe sea ice conditions were favorable for sub-ice sedimentation of the fine-grained diamicton matrix in small freshened basins between still glaciated shoals. Short down-slope suspension flows and seasonal brines cascading were main sedimentation mechanisms during the late deglaciation, although more extensive contour currents were possibly active in the southern Barents Sea. According to micropaleontological data, biological productivity was generally low during the deglaciation likely owing to restricted mixing of the subsurface saline Atlantic water and the surface water freshened by glacier melt-water discharge.

In: Deglaciation
Editor: Marcella Boone

ISBN: 978-1-53612-501-6


*Chapter 1*

# DATING THE LAST DEGLACIATION: GEOCHRONOLOGIC METHODS USED TO ASSIGN TEMPORAL CONSTRAINTS TO THE RETREAT OF LATE PLEISTOCENE ICE SHEETS

***Ken Munyikwa****

Centre for Science, Athabasca University, Athabasca, Alberta, Canada

## ABSTRACT

Assignment of a temporal scale to deglaciation events is elemental to the understanding of processes involved during and subsequent to the retreat of continental ice sheets. Dating the recession of ice sheets, however, is complicated by the fact that ice sheets cease to exist following deglaciation, leaving behind no tangible remnants of the ice itself that can directly be analyzed. Hence, dating exercises essentially aim to place constraints on ice sheet recession by examining proxy

* Corresponding author: Centre for Science, Athabasca University, 1 University Drive, Athabasca, AB T8N1T3, Canada, Email: kenm@athabascau.ca.

records that resulted from or are associated with the former existence of ice sheets. A deglaciation timescale enables rates of postglacial geomorphic processes as well as the pace of environmental evolution to be determined. Additionally, such chronologies facilitate the analysis of variations in the distribution of biotic communities over time, in conjunction with changes in climatic conditions and other aspects of the environment. In areas where isostatic rebound occurred following glacial unloading, accurate dating allows geophysical models of viscoelastic responses to be calibrated, including changes in sea level. This chapter provides a review of dating methods that have been used to constrain the retreat of Late Pleistocene ice sheets from formerly glaciated regions following the Last Glacial Maximum, with a focus on North America. Chronological methods explored include cosmogenic-nuclide exposure, luminescence, radiocarbon and varve dating techniques. In each instance, the principles behind the dating technique and the types of material that can be dated are examined. The chapter concludes with a brief outline of a case study conducted in western Canada to constrain the Late Pleistocene retreat of the Laurentide Ice Sheet from the region.

**Keywords:** deglaciation, dating methods, ice sheet, Late Pleistocene, Quaternary, Laurentide Ice Sheet, ice age

## INTRODUCTION

Previous deglaciation events were associated with diverse changes in earth's atmospheric, physical and biotic environments (e.g., Gibbard and Head, 2009; Clark et al. 2012; Cuffrey et al. 2016). A cogent insight into the mechanisms of such events lends an understanding to earth's response mechanisms to climate fluctuations. The ability to reconstruct past deglaciations, however, rests on the capacity to assign a robust temporal scale to observed features and processes. By virtue of its proximity in time, the last deglaciation is the most studied glacial termination to date, with investigations having begun at least 175 years ago (Stroeven et al. 2016). Several geochronologic techniques can be employed to provide absolute age constraints to processes involved in the retreat of Late Pleistocene ice sheets from terrestrial regions (e.g., Jackson et al. 1997, Dyke et al. 2001, 2003; Balco and Schaefer, 2006; Raukas et al. 2010; Ridge et al. 2012;

Stroeven et al. 2016; Munyikwa et al. 2017). Assigning a chronology to the recession of ice sheets, however, is complicated by the fact that ice sheets cease to exist following deglaciation, leaving behind no tangible vestige that can directly be analyzed. Hence, geochronologic methods primarily aim to place constraints on ice sheet recession by examining proxy records that resulted from or are associated with the former presence of large ice sheets on continental landscapes.

Observable terrestrial effects of large-scale continental glaciation following ice sheet recession can generally be grouped into three main categories. The first of these comprises erosional features that result from the abrasional effects of ice sheets over bedrock on which the glaciers move. Following deglaciation, such eroded surfaces become exposed to the atmosphere and there are dating methods that can be used to assign temporal constraints for the exposure of the surfaces. Surfaces of weathering products that are produced by glacial processes are also included in this category. Ages obtained from dating the surfaces would be commensurate with minimum age constraints for the deglaciation.

The second group of proxy records consists of depositional materials emplaced by glaciers. The materials include sediments deposited directly by melting glaciers, such as glaciofluvial materials, or glacial sediment reworked by eolian processes during the early postglacial period. Sediments that are released by glaciers, if deposited in water bodies such as lakes as annual layers, can also be used to determine ages that constrain the time of deglaciation. In all instances, ages from the depositional features would serve as minimum age constraints for the timing of glacial retreat.

Since glaciated regions are generally not amenable to most faunal and floral organisms, such terrains are generally barren. Hence, the third and final category of proxy indicators of deglaciation includes biotic communities that colonize formerly glaciated regions as climatic conditions ameliorate. By determining the ages of the earliest organic materials to emerge on the postglacial landscape, it may be possible to establish the earliest time by which the organisms would have been able to flourish following the retreat of ice sheets. Hence, the chronologies would be conservative estimates for the timing of the retreat of the ice sheets.

The importance of assigning constraints to ice sheet retreat cannot be overstated. The dating allows ages of postglacial landforms to be ascertained (e.g., Balco and Schaefer, 2006; Munyikwa et al. 2017). It also enables rates of postglacial geomorphic processes as well as rates of environmental evolution to be determined. In areas where isostatic rebound occurred following glacial unloading, accurate dating allows geophysical models of viscoelastic responses to be calibrated (e.g., Tarasov and Peltier, 2006). Furthermore, environmental variables that are ascertained with an effective temporal resolution enable the leads and lags to be differentiated (e.g., Lowe and Walker, 2015), which would be an important distinction when determining forcing factors.

This chapter provides a review of dating methods that have been used to constrain the retreat of Late Pleistocene ice sheets from formerly glaciated regions following the Last Glacial Maximum (LGM). Focus is placed on terrestrial records and, while emphasis will be on North American ice sheets, with particular attention on the Laurentide Ice Sheet (LIS), the dating methods examined should be globally applicable. Notably, the LIS was the largest Late Pleistocene ice sheet in the Northern Hemisphere. Additionally, though it was also the last of the major ice sheets to terminate, it had a major effect on when the earth system transitioned to Holocene conditions (Ullman et al. 2015). Thus, the ice sheet is of major interest to many researchers. This chapter comprises ten sections. Following the introduction in the first section, the Quaternary chronological framework as well as causes of glaciation and deglaciation during the Pleistocene are examined in the second section. In the third section, features of formally glaciated landscapes that can be dated using geochronologic methods are looked at. The fourth section presents an overview of dating methods that can be used to constrain deglaciation before examining specific methods in subsequent sections. In the fifth section, the use of cosmogenic radionuclides for ascertaining exposure ages of formerly glaciated terrains is explored. The sixth section examines the use of luminescence dating methods to constrain depositional ages of sediments on postglacial landscapes. Age constraints based on radiocarbon dating performed on remains of biological organisms, floral or faunal, that colonize postglacial landscapes are reviewed in the seventh section. In the eighth section, varve chronology as well as other dating methods that could be used to date deglaciation are explored. As an illustration, a case study conducted in western Canada to constrain the retreat of the LIS from the region is briefly discussed in the ninth section. The chapter concludes in the tenth section by examining future trends and methodological advances in dating techniques that may expand prospects in dating deglaciation going forward, including dating glacial retreats beyond the last deglaciation.

## Causes of Late Pleistocene Glaciation and Deglaciation

### Chronological Framework of the Quaternary Period

Before examining geochronological methods that can be used to constrain the retreat of Late Pleistocene ice sheets since the LGM, it is imperative to outline the chronological framework used to study the Quaternary Period as well as the driving forces behind climatic fluctuations. The Pleistocene Epoch spanned from ca. 2.58 million years (Ma) to around 11.7 thousand years (ka) ago (Gibbard et al. 2010). Together with the Holocene Epoch that began at the end of the Pleistocene Epoch, ca. 11.7 ka ago (e.g., Walker et al. 2009), and has lasted until the present, they make up the Quaternary Period. The Quaternary stands apart from previous geological periods in that it was characterized by intense cyclical fluctuations during which the global climate experienced intense cold periods. Accompanying the cold periods was the expansion of Northern Hemisphere continental ice sheets (Sarnthein et al. 2009) and mountain glaciers in many regions (Lowe and Walker, 2015). Between the periods of glaciation, the climate ameliorated for shorter intervals referred to as interglacials. The duration of the glacial-interglacial cycles appears to have changed over the course of the Quaternary. Around the middle of the Pleistocene (ca. 800-900 ka ago), the cycle changed from around 41 ka to around 100 ka during what is referred to as the Middle Pleistocene Transition (MPT) (e.g., Head et al. 2008; Paillard and Parrenin, 2004; Lowe and Walker, 2015). The duration of glacials since the MPT have in fact ranged between 80-120 ka. For instance, about 400 ka ago, glacials lasted close to 80 ka, whereas the most recent glacial stage was around 120 ka long. Data reflecting these changes can be obtained from deep-sea oxygen isotope records (e.g., Shackelton and Opdyke, 1973; Shackleton et al. 1990; Broeker and Denton, 1990; Hilgen et al. 2006). Correlable data are also reflected in ice-core records from Greenland (up to the last interglacial) (Dahl-Jensen et al. 2003; Shackleton et al. 2004; Svensson et al. 2008) and from Antarctica (up to ca. 800 ka) (e.g., Jouzel et al. 2007; Masson Delmotte et al. 2010). With regards to driving forces behind the fluctuations, though orbital forcing is thought to be the underlying driver for glacial-interglacial cycles, other factors may be involved in the changes in periodicities. As will be discussed below, these include changes in the global distribution of continental landmasses as well as variations in atmospheric composition (Lowe and Walker, 2015). However, firm answers still elude investigators.

Notably, cold and warm climatic variations are not unique to the Quaternary. Throughout much of the Cenozoic Era and also during other preceding eras, geologic records show that climatic fluctuations between cold and warm phases did occur (Zachos et al. 2001). What makes the Quaternary unique is the greater amplitude as well as the higher frequency of climate variations. Hence, the beginning of the Quaternary signifies a major change in the earth's climate system. Edward Forbes in 1846 referred to the Quaternary as the Glacial Epoch (Lowe and Walker, 2015), and many today refer to the epoch informally as the ice age.

Following a proposal from the Subcommission on Quaternary Stratigraphy (SQS) and the International Union of Quaternary Research (INQUA), the International Commission on Stratigraphy (ICS) approved in 2009 a formal tripartite subdivision of the Pleistocene Epoch into the

Early, Middle and Late Pleistocene subepochs (Gibbard and Head, 2009) (Figure 1). Notably, the terms early, middle and late in reference to the Pleistocene had been in use for some time, albeit informally, and the terms lower and upper have also been used synonymously with early and late, respectively.

Late Cencozoic Time Scale showing ICS approved clasification for the Quaternary

| ERA | PERIOD | EPOCH & SUBEPOCH | | AGE | AGE (Ma) | GSSP |
|---|---|---|---|---|---|---|
| CENOZOIC | QUATERNARY | HOLOCENE | | | | |
| | | PLEISTOCENE | Late | Tarantian | 0.012 | |
| | | | Mid | Ionian | 0.126 | |
| | | | Early | Calabrian | 0.781 | |
| | | | Early | Gelasian | 1.806 | ← Vrica, Calabria |
| | | PLIOCENE | | Piancenzian | 2.588 | ← Monte San Nicola, Sicily |
| | | | | Zanclean | 3.600 | |

Figure 1. Framework of the Quaternary approved by the International Commission on Stratigraphy (ICS) (after Gibbard and Head, 2009). GSSP refers to the Global Boundary Stratotype Section and Point.

While, as of early 2017, this framework is still to be ratified by the International Union of Geological Sciences (IUGS), the subepochs are used in this chapter as formal terms. As approved by the ICS, the Late Pleistocene corresponds with the Tarantian age that began ca. 126 ka ago and terminated at the end of the Pleistocene Epoch around 11.7 ka ago (Gibbard and Head, 2009). The last glaciation event began around 115 ka ago and terminated at the end of the Pleistocene around 11.7 ka ago. Hence, the Late Pleistocene includes the penultimate interglacial (Eemian in Europe) and the last glaciation event that lasted from ca. 110 to 11.7 ka ago.

## Drivers of Climate Change

According to Zachos et al. (2001) global climate change during the Cenozoic (the last 65 Ma) has been driven by three main factors: tectonic processes, orbital forcing and random events. Tectonic processes have forced cooling and warming trends that are of timescales in the $10^2$ ka to $10^4$ ka range. Variations driven by orbital forcing are of the time range 10 ka to $10^3$ ka while aberrant shifts are of the range $10^2$ ka.

The influence of tectonics on climate is usually unidirectional (as opposed to cyclic) and typically affects boundary conditions such as geography of continents, elevation, oceanic outlets and seafloor topography (Zachos et al. 2001). Effects of tectonic events on climate are best illustrated by what is thought to have triggered the onset of the more intense climatic fluctuations at the start of the Quaternary around 2.58 Ma ago. The closure of the Central American Seaway that lay between North and South America around 2.8 Ma ago (O'dea et al. 2016) has been proposed (Weyl, 1968; Keigwin, 1982) to have augmented meridional overturning circulation in the northern Atlantic (Bartoli et al. 2005) that in turn triggered Northern Hemisphere glaciation that is typical of the Quaternary (Sarnthein et al. 2009). The closure is thought to have stemmed from tectonic plate convergence in Central America that led to the subduction of what was then the Farallon Plate under the Caribbean and South American plates. Associated island arc volcanism on the Caribbean Plate resulted in the formation of the Panama Arc. Following further movement, the Panama Arc merged with the South American continent to form the isthmus (O'dea et al. 2016). While many investigators (e.g., Keigwin, 1978; Jackson and O'Dea, 2013; Coates and Stallard, 2013) think that the closure of the seaway occurred around 2.8 Ma, it is noteworthy that other studies have argued that the isthmus formed as early as 4.6 Ma (e.g., Haug and Tiedemann, 1998) while others suggest an age of formation up to 10-11 Ma ago (e.g., Montes et al. 2015). Such timelines would imply that the closure of the Panama isthmus was not the trigger for the onset of the Quaternary. Additional questions about the viability of the Panama isthmus have been raised by a number of other researchers (e.g.,

Ravelo et al. 2004; Molnar, 2008). However, Haug and Tiedemann (1998) suggest that even if the closure of the isthmus had occurred around 4.6 Ma, it would have resulted in ocean circulation changes that promoted Northern Hemisphere warming. This in turn resulted in increased moisture yields in the Northern Hemisphere that, in conjunction with orbital forcing, promoted the growth of Northern Hemisphere ice sheets between 3.1 and 2.5 Ma ago (Haug et al. 2005), which would be consistent with the chronology of the start of the Quaternary.

Alternative tectonic explanations for the onset of the Quaternary include the suggestion by Cane and Molnar (2001) that the closure of the Indonesian seaways around 3-4 Ma by the combined effect of tectonic movement and volcanic activity of Halmahera Island was responsible for the modification of ocean circulation that led to changes in the global earth-climate system (Sarnthein et al. 2009). The uplift of the Tibetan Plateau resulting from the collision between the Indian Plate and Eurasian Plate has also previously been suggested to have ushered in cooler temperatures during the Cenozoic (e.g., Raymo and Ruddiman, 1992; Garzione, 2008). In particular, increased denudation that followed the uplift is thought to have led to the drawdown of atmospheric $CO_2$, ultimately to be sequestered in marine carbonates (e.g., Walker et al. 1981; Berner et al. 1983; Zachos and Kump, 2005; Garzione, 2008). Combined with burial of organic carbon, this is thought to have resulted in lower global temperatures (Garzione, 2008).

Nonetheless, despite the divergent opinions on the exact timing of what might have triggered the onset of the Quaternary, the general consensus today seems to be that tectonic forcing may have played a role, directly or indirectly, in triggering onset of intensified climate fluctuations that have characterized the Quaternary over the last 2.6 Ma.

## Triggers of Glacial Stage Terminations

As mentioned above, processes that trigger deglaciation are not fully understood at present (e.g., Ullman et al. 2105). Nonetheless, deep-sea and ice core records have shown that glacial terminations occur relatively rapidly and orbital influence alone cannot account for the rapid change (e.g., Denton et al. 2010; Lowe and Walker, 2015). Using high resolution temporal records, a number of divergent mechanism have been proposed for the termination. The hypotheses demonstrate that the terminations are thought to be the product of the interplay between variations in orbital insolation resulting from astronomical forcing as well as response processes from earth systems. One suggestion put forward (Stott et al. 2007; Lowe and Walker, 2015) is that, since data from Antarctica indicate that warming there started just after the LGM but slightly earlier than in the Northern Hemisphere, events in the Southern Hemisphere may have triggered the termination (see also Paillard and Parrenin, 2004). Atmospheric $CO_2$ levels obtained in ice cores also lagged behind the heating noted in the Southern Hemisphere, which would imply that $CO_2$ rise was not the trigger for the deglaciation (e.g., Timmerman et al. 2009). As a result, some investigators have proposed that the higher temperature in the Southern Hemisphere led to the retreat of sea ice in Antarctica. This would have led to increased ventilation in the Southern Ocean resulting in higher atmospheric $CO_2$ levels (Stott et al. 2007; Lowe and Walker, 2015).

Other researchers, however, have suggested that the last deglaciation began with events in the Northern Hemisphere. It is thought that towards the end of the LGM, with ice sheets at their maximum, stadial events in Greenland had become longer. As a result of the bipolar seesaw (whereby cooler periods in the north caused warmer periods in the south) (e.g., Barker et al. 2009; Severinghaus, 2009), warmer periods were noticed in the Southern Hemisphere during the prolonged stadials in the north. Increased ventilation in the south caused by the warming also led to a greater release of $CO_2$ to the atmosphere which in turn amplified the warming. Cumulatively, these effects would have swung the climatic pendulum in the north in such a direction that a return to sustained cool conditions was not possible. Northern monsoon data reported by Cheng et al. (2009) appear to be consistent with this scenario. These suggest that periods during which ice sheets waned in the Northern Hemisphere coincided with phases of higher sea level, weakened monsoons and higher $CO_2$ levels in the atmosphere.

An alternative approach has been to look at the chronology of temperature change between ca. 22 and 5 ka which has shown that the first significant spike in temperature was in the Northern Hemisphere around 19 ka (He at al., 2013). Subsequently, the temperature in the Southern Hemisphere rose above that in the north around 18 ka, and between 18 ka and the start of the Holocene, the temperature in the north lagged behind the south. This is thought to indicate that processes in the Northern Hemisphere initiated deglaciation (Shakun et al. 2012; He et al. 2013). The process started with a warming event in the Northern Hemisphere due an increase in summer insolation around 22 ka. This would have caused ice sheets to melt, lowering the albedo and increasing temperature in Greenland (Sevenringhaus, 2009; Shakun et al. 2012; He et al. 2013). Freshwater released by the melting glaciers would then have shut the Atlantic Meridional Overturning Circulation (AMOC) to result in a cooling of the North Atlantic. With the AMOC not running, the Southern Ocean and Antarctica became warmer and $CO_2$ release into the atmosphere in turn augmented the heating (Shakun et al. 2012). Over time, the heating spread beyond the Southern Hemisphere, causing additional melting of ice sheets in the Northern Hemisphere. Released water in the northern Atlantic may have helped to keep the AMOC inoperative such that the north remained cool. Eventually, nonetheless, heating became global, leading to the termination of the last glacial (Severinghaus, 2009; He et al. 2013; Lowe and Walker, 2015). A relatively similar mechanism for the Late Pleistocene termination was proposed by Sigman et al. (2007).

Thus, while no definitive process for the initiation of the last deglaciation has been ascertained, high resolution records from deep sea sediments and ice-core data from both Greenland and Antarctica have allowed researchers to propose highly plausible hypotheses of what might have taken place. What is indisputable, is that climate records demonstrate that deglaciation began around 21.5 ka and ended around 11 ka. Hence, as stated above, this chapter focuses on methods used to constrain the timing of ice-sheet retreat from continental regions between the start of glacial termination and the ultimate disappearance of the ice sheets.

## Features of Previously Glaciated Landscapes: What Can be Dated?

By virtue of their magnitude, ice sheets leave indelible signatures of their presence on any landscape over which they form or move. If temporal constraints can be assigned to such features, it would enable approximations to be made for when the ice sheets existed or disappeared from the landscape. This section examines features of former glaciated landscapes that lend themselves to dating using geochronological methods.

### Erosional Landforms

Glacial landforms can generally be classified into two main categories: erosional landforms and depositional landforms. Glacial erosion occurs mainly through processes of abrasion and plucking. Because ice itself is relatively soft (Mohs hardness of about 1.5) compared to rock, abrasion occurs through the scraping and gouging of rock embedded in the ice at the base of the ice sheet. This abrasive action can in turn generate large volumes of fine grained rock, sometimes referred to as rock flour. When the rock flour is trapped between the base of the glacier and the rock surface, the movement of the glacier causes the flour to act as an abrasive that polishes the rock surface further (Smith and Pun, 2010). Important for dating purposes, as will be discussed below, abrasion leaves behind freshly denuded rock surfaces that become subaerially exposed once the ice melts.

Glacier erosion that occurs through plucking takes place along preexisting weaknesses within rocks such as joints and faults which allow pieces of loose fragments to be extracted from the main rock body. Plucking is particularly effective at the base of the glacier because the shear stress is highest there. Freeze-thaw action also operates at the bottom of the glacier but is not limited to the base. Some freeze-thaw activity occurs at high elevation above the glacier which results in rock falls onto the glacier where they get transported by the moving ice. Large boulders along the Erratics Train in western Canada are thought to have been transported in this manner (e.g., Jackson et al. 1997; Hartman and Clague, 2008).

An important element that needs to be established during any dating exercise that aims to constrain deglaciation is the direction of ice movement and erosional structures play an important role in determining the direction. Features that can be used for determining direction of ice movement can be found at both small scale and large scales. Small scale features include striations, polishing and grooves that may appear on rock that protrudes from formerly glaciated surfaces. Striations are formed when rocks embedded into the base of the glacier are dragged across bedrock and, by mapping the striations over a large area, the direction of ice flow can easily be ascertained.

Large scale erosional features include streamlined landforms such as whalebacks, roches moutonnées and glaciated valleys.

## Depositional Landforms

As glaciers move down-gradient, they carry with them vast quantities of material that has been eroded. The products of weathering and erosion are carried as embedded debris within the ice, at the base of the ice as well as on the sides of the ice sheet. Material can also be carried atop the ice sheet. Meltwater that flows at the base of the glacier carries with it large amounts of finer grained sediment. At some point the deposits transported by the glaciers are ultimately emplaced as sediment and these are generally grouped into two main categories: till and outwash. Till refers to sediment that is directly deposited by the glaciers whereas outwash is sediment that is carried by glacial meltwater and is usually deposited some distance from the glacier itself. Typically, till is very poorly sorted and can contain material ranging from boulders to gravel, sand, silt and clay. Till also generally contains no bedding. The provenance of till can include all areas over which the glacier travels and large boulders can be transported over distances exceeding 1000 kilometers to be deposited as erratics. For example, the Prairie provinces of western Canada as well as the Midwestern states of the US are littered with erratics transported from the Canadian Shield located near Hudson Bay to the north.

Also, associated with glaciers are large glacial lakes that form when glacial ice down-gradient dams meltwaters to form extensive bodies of water. Sediment comprising fine sands, silt and clay can be deposited in these lakes. Large deltas can also form where rivers debouch into the glacial lakes. When glacial lakes finally drain away, as deglaciation continues, the former lakes leave behind large quantities of fine-grained sediment subaerially exposed, often on landscapes that bear little vegetation following the glacial phase. As will be discussed in detail later, such deposits are susceptible to eolian mobilization and have recently become a very important target for researchers trying to constrain the timing of ice sheet recession. Table 1 presents characteristic landforms that can be found on formerly glaciated terrains.

Typically, with ice sheets, depositional landscapes are characterized by extensive areas that are covered by thick sequences of ground moraine. Several end moraines representing various recessive positions of the glacier as it recedes can be found on the postglacial surface. Downstream of till, large plains can be covered by outwash deposits.

Another class of landforms that develop on postglacial landscapes and are important to dating are lakes. In areas that were formerly covered by the glaciers, deep depressions may have been scoured into the bedrock and these may fill up with water once the glaciers retreat. An example are the Great Lakes of North America. Another type of lake forms when large bodies of ice that remain following ice sheet retreat are covered or surrounded by outwash deposits. When the ice subsequently melts with ameliorating climate, it leaves behind deep depressions in the outwash deposits that get filled with water to form kettle lakes.

**Table 1. Common glacial landforms**

| Landform | Appearance of landform |
|---|---|
| **Erosional** | |
| **Arete** | Narrow, sharp-edged ridges that separate glacial valleys and cirques. |
| **Cirque** | Hemispherical depression on a mountain side at the rear end of a glacial valley. |
| **Horn** | Steep peaks that form surrounded by three or more cirques. |
| **Roche Moutonée** | Elongated mound of bedrock that has been carved by a glacier aligned with the direction of flow. The mound slopes gently in the direction of glacial movement. |
| **Tarn** | A lake that forms in a deepened cirque. |
| **Depositional Features** | |
| **Drumlin** | A ridge of glacigenic sediment (till or outwash) that is shaped by flow at the bottom of a glacier. The ridge has a concave slope that is steeper in the upstream direction of the glacier and gentler downstream. |
| **Esker** | A sinuous ridge with steep sides that was deposited in a tunnel under a glacier by meltwater and stands out once the ice melts away. |
| **Kame** | Alluvium deposited by glacial meltwaters at the side or under a glacier that stands out as a ridge once the glacier melts away. |
| **Moraine** | Unstratified sediment emplaced at the fore, side or bottom of a glacier. |
| **Sandur** | A plain formed by sediment deposited by meltwater from a glacier. |

## THE LAST DEGLACIATION – APPLICABLE DATING METHODS: AN OVERVIEW

When selecting a dating method, there are two main factors that should be considered. The first is the dating range under investigation. Geochronologic methods have different time ranges over which they can be applied, depending on the principles of the dating method. Thus, cognizance is required of the approximate age range of the events that are being constrained. In the case of the last deglaciation, ice sheets in the Northern Hemisphere were at their Late Pleistocene maximum limits sometime between 26 and 19 ka (Clark et al. 2012). For the LIS, for instance, the eastern and southeastern margins reached their maximum limit around 24 ka (e.g., Balco and Schaefer, 2006) whereas on the western margin, the ice sheet attained its maximum extent around 21 ka (Dyke et al. 2003). We also know that the last deglaciation began around 19-20 ka (Clark et al. 2009) and the onset of the Holocene ca. 11.7 ka signaled the end of the deglaciation process. Therefore dating methods that can be used to constrain the last deglaciation are those that can date the Late Pleistocene period.

Different dating methods usually analyze different materials. Hence, the second consideration that needs to be made when choosing a dating method is to check whether material exists that can be dated using the selected method. Above, it was pointed out that when dating deglaciation, three main categories of materials on the postglacial landform can be targeted for analysis. These include products of erosion, products of deposition as well as organic materials that proliferate on a postglacial landform once the ice sheet retreats. In this chapter, dating methods used to analyze these materials will be placed into four main groups. For the class of materials that includes products of erosion, freshly exposed rock surfaces can be dated using the accumulation of cosmogenic nuclides. Thus, the first dating method to be examined will be a technique that provides time that has lapsed since a given rock surface was exposed to cosmic rays from outer space. The utility of cosmogenic nuclide dating lies in the range of nuclides that can be used for analysis and these enable most rock forming minerals to be employed in the analysis. Consequently, most rock types can be used for cosmogenic radionuclide dating (Ivy-Ochs and Kober, 2008).

The second category of materials belongs to products of deposition by ice sheets. This includes sediments that were reworked from glacial deposits and redeposited either as fluvial sediment or eolian deposits. Luminescence dating, the second group of dating methods discussed, is a Quaternary dating technique that can be used to approximate depositional ages of clastic sediments that contain minerals such as quartz or feldspar.

In order to use ages from such sediments as constraints for deglaciation, deposits that were emplaced immediately after the retreat of the ice sheets are targeted. Both quartz and feldspar are ubiquitous in glacial sediments as well as in sediments sourced from glacigenic deposits because of their resistance to both physical and chemical weathering. Hence, luminescence dating is an approach that has been employed extensively in efforts to constrain the timing of deglaciation.

The third class of materials comprises remains of floral or faunal organisms that colonize postglacial landscapes in the immediate aftermath of the retreat of the ice sheet. Such materials can be dated using radiocarbon methods, which are the third group of dating techniques examined in this chapter. Because the organic material would need to fulfill certain criteria in order to produce reliable constraints for deglaciation, the discussion below will highlight conditions that are necessary for the production of useful data.

The fourth and final group of dating methods examined comprises other methods that have been used to constrain the timing of deglaciation in some settings including varve chronology, which is dating conducted by counting annual laminations in lake sediment (e.g., Balco and Schaefer, 2006). Varves can be found in areas where thick sequences of sediment accumulated in glacial lakes that formed on the periphery of ice sheets during the last deglaciation. Because of the simplicity and yet robust characteristics of the approach, varve chronology was one of the first credible attempts at constructing absolute age chronological frameworks for Late Pleistocene deglaciation (e.g., De Geer, 1912; Antevs, 1922).

In examining the methods described above, the principles behind each dating method are explored after which examples of settings where the dating methods have been used are outlined. The examples will by no means be exhaustive. Rather, these are provided to allow readers who may wish to learn more about the approach to access additional information. Also for each method that is described, sources of errors and problems associated with the technique are outlined briefly.

## SURFACE EXPOSURE DATING USING COSMOGENIC NUCLIDES

Time that has elapsed since fresh rock surfaces became subaerially exposed can be approximated using cosmogenic nuclide dating methods (e.g., Gosse and Phillips, 2001; Ivy-Ochs and Kober, 2008; Dunai, 2010; Balco, 2011; Darvill, 2013; Granger et al. 2013). This is because when silicate minerals are exposed to cosmic rays at the earth's surface, there is an accumulation of some rare nuclides over time in a manner that can be predicted (e.g., Ivy-Ochs and Kober, 2008). Cosmic rays from outer space comprise charged particles such as atomic nuclei as well as subatomic particles that include protons, alpha particles, single electrons and some particles of antimatter (Dunai, 2010). Though some cosmic rays are generated within the solar system, they are of lower energy and those that are of relevance to cosmogenic nuclide dating are galactic cosmic rays that originate from outside the solar system. (Masarik and Reedy, 1995). These travel with high energy towards earth from all directions. The energy and flux of the cosmic rays reaching earth are thought to be approximately constant over the last 10 Ma (Dunai, 2010). Cosmic rays that reach the upper part of earth's atmosphere have protons constituting about 87% of the rays, while alpha particles make up about 12%. Larger nuclei only make up about 1% (Dunai, 2010). The kinetic energy of a large proportion of the primary particles falls between 100 -1000 MeV per nucleon. After entering earth's atmosphere, the cosmic ray particles interact with gaseous particles resulting in the release of a cascade of secondary particles (Figure 2). Secondary particles are largely produced by the process of spallation whereby constituent particles are expelled from their parent gaseous nuclei following collision. Spallation occurs relatively easily because, by comparison, binding energies in the nuclei generally fall in the range 7-9 MeV per nucleon which is less than the kinetic energy of primary cosmic ray particles. The secondary particles may be involved in additional reactions with decreasing distance towards earth but by the time they reach the solid earth, the cosmic rays comprise about 98% neutrons (Masarik and

Beer, 1999). This is largely because protons, which constitute the predominant cosmic ray component in the upper atmosphere, also experience losses through ionization whereas neutrons, which are produced by spallation, do not (Lal and Peters, 1967).

When particles from primary and secondary cosmic rays interact with atomic nuclei, they produce cosmogenic nuclides. At the earth's surface, secondary cosmic ray particles such as neutrons and muons are responsible for the overwhelming majority of cosmogenic nuclides that are produced (Masarik and Beer, 1999). The type of nuclear reaction that occurs depends on the energy of the particles (Dunai, 2010). Spallation, thermal neutron capture and negative muon capture are all processes that can result in the production of nuclides that can be used in earth science studies (Dunai, 2010) (Table 2). Spallation occurs when neutrons collide with atomic nuclei, knocking off protons and neutrons, resulting in a lighter nucleus. Alpha particles can also be emitted. Most of the neutrons that are produced by the interaction between cosmic rays and atomic nucleons in the atmosphere are ultimately retarded by the time they reach the earth's surface such that they become thermal neutrons with energy in the range 0-1eV. Such slow moving electrons may become captured to produce some cosmogenic nuclides. Negative muon capture occurs when negative muons that have been slowed down are incorporated into electron shells of atoms. Once captured by the electron shell, the muons could ultimately end up being incorporated into the nucleus. In the nucleus, a proton is neutralized by the muon and the process delivers about 106 MeV to the nucleus. Emission of alpha particles and nuclei $^{2}H$, $^{3}H$, and $^{3}He$ may also occur (Table 2).

Cosmogenic nuclides most currently used in dating applications are outlined in Table 3. Isotopes $^{10}Be$, $^{14}C$, $^{26}Al$ and $^{36}Cl$ are radionuclides whereas $^{3}He$ and $^{21}Ne$ are stable. Nuclides such as $^{53}Mn$ (Table 2) still have to be developed for routine application. Advantages and drawbacks of each nuclide are also indicated. It is important to note that having reached ground level, the cosmic ray flux falls rapidly with depth. Consequently, the production of cosmogenic nuclides also decreases with depth. It is also pertinent to point out that most of the nuclides commonly used for exposure dating are produced in spallation processes between primary or secondary cosmic ray neutrons and targeted atoms in earth materials (Table 2).

**Table 2. Common cosmogenic nuclides and their production mechanisms (after Dunai, 2010)**

| Isotope | Primary spallation reactions | Thermal neutron capture | Negative muon capture |
|---|---|---|---|
| $^{3}He$<br>$^{3}H(\beta^{-3}He)$ | On elements heavier than particle, typical reactions include: $(n, x^{3}He)$ or $(n, x^{3}H)$, x being any other particle or none | $^{6}Li(n, \alpha)\ ^{3}H$ $(\beta^{-3}He)$ | On elements heavier than particle, typical reaction include: $(\mu^{-}, ^{3}He)$ or $(\mu^{-}, ^{3}H)$ |
| $^{10}Be$ | $^{16}O(n, ^{3}He\alpha)\ ^{10}Be$ or | $^{9}Be(n, \gamma)\ ^{10}Be$ | $^{16}O(\mu^{-}, \alpha pn)\ ^{10}Be$ |
| | $^{16}O(n, 4p3n)\ ^{10}Be$ | | $^{28}Si(\mu^{-}, x)\ ^{10}Be$ |
| | $^{28}Si(n, x)^{10}Be$ | | |
| $^{14}C$ | $^{16}O(n, 2pn)\ ^{14}C$ | $^{17}O(n, \alpha)\ ^{14}C$ | $^{16}O(\mu^{-}, pn)\ ^{14}C$ |
| | $^{28}Si(n, x)\ ^{14}C$ | $^{14}N(n, p)\ ^{14}C$ | |
| $^{21}Ne$ | $^{23}Na(n, ^{3}H)\ ^{21}Ne$ | | $^{23}Na(\mu^{-}, 2n)\ ^{21}Ne$ |
| | $^{24}Mg(n, \alpha)\ ^{21}Ne$ | | $^{24}Mg(\mu^{-}, ^{3}H)\ ^{21}Ne$ |
| | $^{27}Al\ (n, \alpha^{3}H)\ ^{21}Ne$ | | $^{27}Al(\mu, \alpha 2n)\ ^{21}Ne$ |
| | $^{28}Si\ (n, 2\alpha)\ ^{21}Ne$ | | |
| $^{22}Ne$ | $^{23}Na(n, pn)\ ^{22}Ne$ | | $^{23}Na(\mu^{-}, n)\ ^{22}Ne$ |
| | $^{25}Mg(n, \alpha)\ ^{22}Ne$ | | $^{24}Mg(\mu^{-}, pn)\ ^{22}Ne$ |
| | $^{27}Al\ (n, \alpha pn)\ ^{22}Ne$ | | $^{27}Al(\mu^{-}, \alpha n)\ ^{22}Ne$ |
| | $^{29}Si\ (n, 2\alpha)\ ^{22}Ne$ | | |
| $^{26}Al$ | $^{28}Si(n, 2np)\ ^{26}Al$ | | $^{28}Si(\mu^{-}, 2n)\ ^{26}Al$ |
| $^{36}Cl$ | $^{39}K(n, \alpha)\ ^{36}Cl$ | $^{35}Cl(n, \gamma)\ ^{36}Cl$ | $^{39}K(\mu^{-}, ^{3}H)\ ^{36}Cl$ |
| | $^{40}Ca(n, \alpha n)\ ^{36}Cl$ | $^{39}K(n, \alpha)\ ^{36}Cl$ | $^{40}Ca(\mu^{-}, \alpha)\ ^{36}Cl$ |
| $^{53}Mn$ | $^{54}Fe(n, pn)\ ^{53}Mn$ | | $^{54}Fe(\mu^{-}, n)\ ^{53}Mn$ |
| | $^{56}Fe(n, 3np)\ ^{53}Mn$ | | |

**Table 3. Cosmogenic nuclides commonly used for dating at present (after Ivy-Ochs and Kober, 2008)**

| Nuclide | Half-life | Analytic. method | Prod. Rate in at. $g^{-1}yr^{-1}$ | Advantages | Drawbacks |
|---|---|---|---|---|---|
| $^{10}Be$ | 1.51 Ma | AMS | 5 | From quartz which is ubiquitous and resistant. | Low production rate; restricted source (quartz); interference in AMS from $^{10}Be$. |
| $^{26}Al$ | 716 ka | AMS | 31 | From quartz; ubiquitous, high production rate. | Restricted to quartz; accurate determination of $^{27}Al$ required. |
| $^{36}Cl$ | 301 ka | AMS | Depends on rock type | Low detection limit; any rock type, silicates, carbonates. | Complicated production; interference from $^{36}S$ in AMS; rock and total Cl composition required. |
| $^{14}C$ | 5.73 ka | AMS | 16 | Convenient for short time scales; quartz ubiquitous. | Short half-life; contamination from atmospheric $^{14}C$. |
| $^{3}He$ | stable | Static MS | 120 | High production rate; good for long timescales; pyroxene, olivine. | Diffuses out of quartz or volcanic groundmass; correction required. |
| $^{21}Ne$ | stable | Static MS | 20 | Useful for long timescales (> 50ka); quartz, olivine, pyroxene. | Correction for magmatic or nucleogenic sources; elevated atmospheric background. |

Abbreviations in table: Analytic. Method – measurement method; Prod Rate in at. $gr^{-1}$ $yr^{-}$– Production rate in atoms per gram, per year. AMS – Accelerator Mass Spectrometry; Static MS – Static Mass Spectrometry.

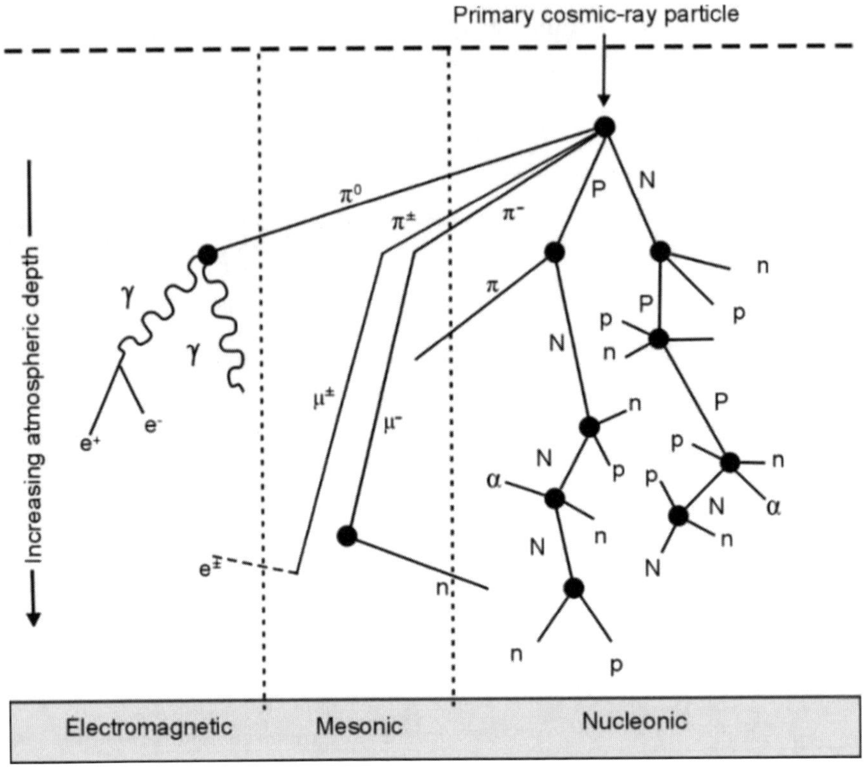


Figure 2. The generation of a secondary particle cascade in the atmosphere by incoming primary cosmic ray particles (adapted from Dunai (2010) and Darvill (2013)). The secondary particles comprise protons (P, p), neutrons (N, n), alpha particles ($\alpha$) and mesons including pions ($\pi$) and muons ($\mu$) (upper case letters denote particles propagating the nuclear cascade). Electromagnetic particles include electrons or positrons ($e^{\pm}$) and gamma-ray photons ($\gamma$).

## Determining Age Using Cosmogenic Nuclides

Determining an age using cosmogenic nuclides requires the evaluation of two main parameters: the production rate of nuclides at the study site as well as the concentration of the nuclide that has accumulated in the surface being analyzed since the beginning of the exposure. Methods by which these variables are ascertained are outlined below.

### *Determining the Nuclide Production Rate*

The rate at which nuclides are produced at the surface of earth depends on a number of factors. One is the effect of the earth's magnetic field on cosmic rays of galactic origin. The activity of the sun also influences the cosmic rays but to a much lesser degree (e.g., Lifton et al. 2005). The extent to which incoming cosmic rays are deflected by the earth's magnetic field is measured by a term referred to as the rigidity. Cosmic rays must exceed the cut-off rigidity to pass through the atmosphere and since the intensity of the earth's magnetic field varies with latitude, the cut-off rigidity also varies with distance from the equator where it peaks and decreases towards the poles (Dunai, 2010).

Another important consideration is that the cosmic radiation flux at the surface of earth changes over time as a result of variations in the earth's magnetic field. These changes take place at timescales of 1 ka -100 ka and need to be taken into account (Dunai, 2010). A third aspect that influences cosmogenic nuclide generation at the earth's surface is altitude, or atmospheric depth to be more precise. This is because the nuclide production rate varies with changes in atmospheric pressure or density of material through which the secondary particles pass and is referred to as attenuation. Hence, variations in atmospheric pressure over time are also important to consider (e.g., Staiger et al. 2007; Dunai and Lifton, 2014).

Production rates of nuclides are usually determined by measuring concentrations of given nuclides in rocks whose exposure age is well known after which these are used to model production rates for other sites at different localities that have different exposure histories (Lifton et al. 2014). This practice, referred to as scaling, entails accounting for the effects of geomagnetic changes over latitude and time as well as the influence of atmospheric depth and other factors to arrive at a production rate that is representative. At present there are two main types of scaling models that are in use (e.g., Lai et al. 1991; Lifton et al. 2005) but there are discrepancies between their results. Hence, concerted efforts continue to find scaling models that would have universal acceptance (e.g., Lifton et al. 2014). More details on scaling models can be found in Dunai (2010), Dunai and Lifton (2014) and Lifton et al. (2014).

***Measuring Nuclide Concentration and Sample Collection***

Though the possibility of using cosmogenic nuclides for dating was noted early (e.g., Davis and Schaeffer, 1955), their application only became possible in the 1970s after the development of more sensitive mass spectrometers that could measure noble gasses as well as the development of accelerator mass spectrometry (AMS). These enable nuclides of low concentration to be measured. Hence today, nuclide concentration measurements for exposure dating purposes are conducted using AMS, in the case of radionuclides $^{10}Be$, $^{26}Al$ and $^{36}Cl$, and noble gas mass spectrometers in the case of $^{3}He$ and $^{22}Ne$.

To collect samples in the field, around 500 g to 1 kg of rock sample is collected using a saw or a drill to extract the top few centimeters from the rock surface. A hammer and chisel could also be used for the same purpose. At any given site, collecting several samples from different boulders or surfaces is preferred. If exposure ages are to be representative, the surface being dated should have remained in the same position since the initial exposure. Hence, large boulders that are immobile are preferred. The surface should not have been covered by detrital material such that the exposure was uninterrupted. At most, only minimal weathering should have occurred in order to minimize loss of the cosmogenic nuclides. Furthermore, pointed features and surfaces close to edges should be avoided (e.g., Masarik and Wieler, 2003; Ivy-Ochs and Kober, 2008).

## What Can Be Dated Using Cosmogenic Nuclides?

As indicated above, the wide variety of nuclides that can be analyzed allows the technique to be applied to any rock type. Consequently, the technique can be used to get exposure ages of bedrock surfaces that have been polished by fluvial processes, glaciers or ice sheets. Similarly, the technique can be used on footwall surfaces from tectonic faults or surfaces exposed following landslides (Ivy Ochs and Kober, 2008). The ages of transported sedimentary deposits such as moraines, mass wasting sediments, fluvial deposits and debris flows can also be approximated by dating surfaces of boulders contained in the sediment.

Cosmogenic nuclide dating has some advantages when compared to other dating methods that can date materials of Late Pleistocene age such as radiocarbon dating. First, with cosmogenic nuclide dating, the surface being analyzed can be dated directly as opposed to obtaining ages that bracket the feature being studied (Darvill, 2013) or to dating other materials associated with the surface. Second, organic material is often transported before being deposited such that contemporaneity between a host deposit and the organic material could be in question. Third, organic material is prone to contamination by older or younger carbon which would yield erroneous ages. Fourth, radiocarbon requires organic material to be present in a sample before it can be dated. This can be a limitation in glacial landscapes where organic materials are generally rare. None of these hindrances apply to cosmogenic methods, which allows for a broader range of materials to be dated. As will be noted in the next section, inadequate bleaching, which is a problem that could affect luminescence dating samples, is also not a problem with dating sediments using cosmogenic methods. The dating age range that can be attained using cosmogenic dating methods depends on the nuclide in question and this varies with geological situations as well as the sensitivity of the measurement technique. Overall, however, cosmogenic nuclide techniques allow dating from a few hundred years to over 10 Ma, which is a much wider chronological range than what can be attained using either radiocarbon or luminescence methods.

## Constraining the Last Deglaciation Using Cosmogenic Nuclides

As outlined above, the time of deglaciation of a given landscape can be determined by analyzing bedrock surfaces that were eroded by glaciers and exposed following the retreat of the ice sheets. Surfaces of boulders that were deposited directly as moraine by the ice sheets as they retreated can also be dated.

***Analysis of Bedrock Surfaces***

The timing of the retreat of ice from a bedrock surface that has been eroded by a glacier (or ice sheet) can be estimated using cosmogenic nuclide dating. For this approach to produce representative results, however, at least three meters of the bedrock cover should be removed before the denuded surface is exposed to cosmic rays. If this does not occur, the freshly exposed surface could contain nuclides from a previous period of exposure which would lead to erroneous age determinations (e.g., Fabel et al. 1997; Colgan et al. 2002). Such inheritance of residual nuclides could occur if the rock is highly resistant to abrasion (Guido et al. 2007). Roches moutonnées are examples of bedrock features that could be targeted for sampling when dating glacier-modified bedrock as the abrasion is often adequate (Ivy-Ochs and Kober, 2008).

The accumulation of cosmogenic nuclides ($C_{(t)}$) in a rock surface that is exposed can be approximated in units of atoms g$^{-1}$ using the equation:

$$C_{(t)} = \frac{P_{(0)}}{\lambda+\frac{\rho\varepsilon}{\Lambda}}\left(1 - e^{-\left(\lambda+\frac{\rho\varepsilon}{\Lambda}\right)t}\right) + C_{in}e^{-\lambda t} \qquad (1)$$

where t is time in *anna* (a) (years) or the exposure age of the surface being investigated, $P_{(0)}$ is the cosmogenic nuclide rate of production at the sampling site in atoms g$^{-1}$ a$^{-1}$ (or atoms per gram, per year), $\lambda$ (in a$^{-1}$) is the decay constant, $\rho$ (in g cm$^{-3}$) is the density of material being irradiated, $\varepsilon$ (in cm a$^{-1}$ ) is the rate of erosion, and $\Lambda$ (in g cm$^{-2}$) is the attenuation length (Ivy-Ochs and Kober, 2008). The concentration of nuclides present at the start of the exposure or the inheritance is represented by $C_{in}$ (in atoms g$^{-1}$). Inheritance of nuclides results in ages that are older than the real age of exposure.

The increase of radionuclides in minerals that are exposed to cosmic rays continues until the number of nuclides produced matches the number that are decaying. This is referred to as secular equilibrium. Erosion from a rock surface in combination with decay results in secular equilibrium being reached sooner than otherwise (Ivy-Ochs and Kober, 2008).

At a site where no erosion is taking place and where glacial abrasion has removed all inherited nuclides, Equation 1 simplifies to:

$$C_{(t)} = \frac{P_{(0)}}{\lambda}\left(1 - e^{-\lambda t}\right) \tag{2}$$

To calculate the age since exposure, Equation 2 is solved for *t*. Scaling factors (accounting for latitude, atmospheric depth and temporal variation in earth's magnetic field as outlined above), sample thickness and topographic shielding would need to be taken into account in order to determine the nuclide production rate ($P_{(0)}$). For the period since the LGM (ca. 20-24 ka), the concentration of nuclides measured in a sample should be directly proportional to the period of exposure (Figure 3).

If the rock surface has experienced erosion since the exposure event that is being dated, Equation 1 would need to be modified accordingly. Otherwise, ages obtained would be younger than the true age. Additional examples of scenarios that can be encountered in the calculation of expoure ages can be found in more comprehensive reviews (e.g., Ivy-Ochs and Kober, 2008).

In situations where the cosmogenic nuclide contribution from a previous glacial event can be determined following limited abrasion, it can be possible to approximate the degree of erosion that occurred during a given glacial event (e.g., Davis et al. 1999; Colgan et al. 2002). This would in essence allow one to calculate the rate of glacial downwasting (e.g., Fabel et al. 2004). To make such a determination, one would need to evaluate the concentration of cosmogenic nuclides that accrued in the rock surface before the glacial event as well as concentration in the rock after the deglaciation. Experimental design and methods for calculating rates of glacial erosion have been outlined by Fabel et al. (2004).

Studies where bedrock abrasion has been dated to constrain deglaciation include an investigation by Gosse et al. (1995) who used $^{10}Be$ from roches moutonnées in western Wyoming to constrain the duration and retreat of Late Pleistocene glaciers in the Rocky Mountains. The postglacial retreat of valley glaciers in the Rocky Mountains in Colorado has been investigated by Guido et al. (2007) who used $^{10}Be$ sampled from polished bedrock. Akçar et al. (2008) used $^{10}Be$ to date surfaces that included samples from roches moutonnées as well as from glacially abraded bedrock to constrain the retreat of a glacier from a mountain valley in Turkey. In the central Swiss Alps, Kelly et al. (2006) used cosmogenic $^{10}Be$ to constrain the deglaciation of the Grimsel Pass region. Additional examples of studies that employed exposure ages from abraded surfaces can be found in Balco (2011).

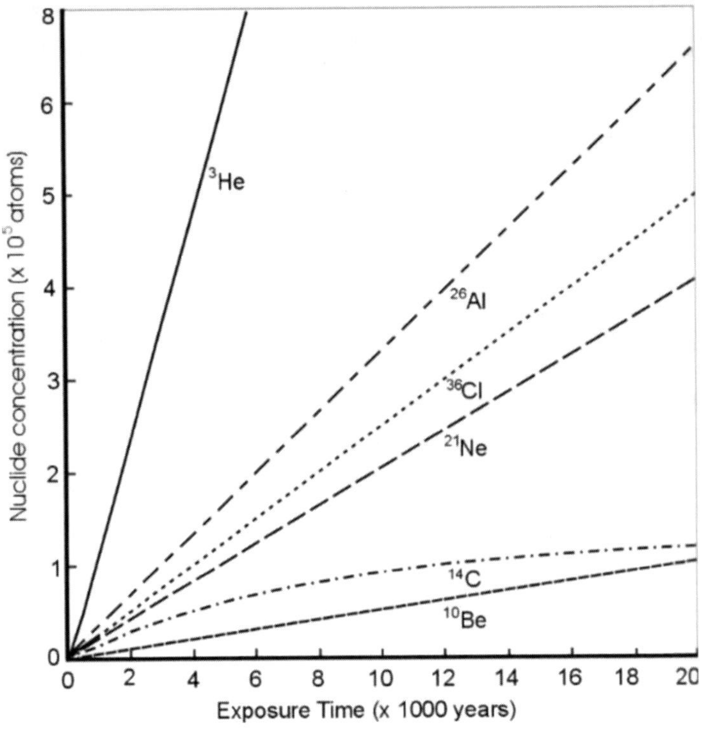


Figure 3. Increase in concentration of cosmogenic nuclides in an exposed rock surface over the last 20 ka (after Ivy-Ochs and Kober, 2008). This shows that for late Quaternary dating purposes, a period encompassing the last deglaciation, there is a direct relationship between time and cosmogenic nuclide concentration.

It deserves to be mentioned that when dating glacially abraded bedrock surfaces, it has been noted that the erosion is often inadequate to erase previous accumulations of cosmogenic nuclides (e.g., Briner, 1998; Bierman et al. 1999) which, as indicated above, leads to inheritance. A complicating factor is that it is not possible to know in advance if abrasion completely reset an exposed surface. As a result, investigators now tend to favor dating boulders that occur in moraine as these appear to yield better results (Balco, 2011). It has also been noted that surfaces eroded by plucking, as opposed to abrading, tend to have less problems related to inheritance. Hence, these too are targeted whenever possible (Balco, 2011).

***Dating Moraines***

The utility of moraines in placing temporal constraints on deglaciation lies in the fact that they delineate the margin of the ice sheet or glacier at a given time from the past. Moraines furthest from the point of final termination would indicate the margin of the ice sheet at the start of deglaciation. When dating moraine, large boulders in stable positions along the summit of the moraine are usually preferred. This is because, as metioned earlier, exposed surfaces should have remained exposed since the retreat of the glacier and also that the sample surface should be free of any obtructions or shielding to cosmic rays exposure (Putkonen and O'neal, 2006). Possible complicating factors could include surface weathering of the boulders and disintegration of the moraine. Another important aspect to note when dating boulders is that depending on which part of the original bedrock the boulder came from, it could have some inheritance that would need to be taken into account when determining the age of exposure. However, Putkonen and Swanson (2003) have reviewed a number of studies and observed that inheritance in boulders was only encountered in a minority of cases.

The first large scale study that employed surface exposure dating using cosmogenic nuclides was conducted by Phillips et al. (1990) on the Bloody Canyon moraine in the Sierra Nevada using $^{36}Cl$. Since then, hundreds of other studies have been conducted worldwide and a comprehensive review has been provided by Balco (2011). Some of these include a study by Balco and Schaefer (2006) who used $^{10}Be$ to ascertain chronologies for the Late Pleistocene deglaciation of southern New England. In western Canada, Jackson et al. (1997) used $^{36}Cl$ on boulders from the Foothills Erratics Train to constrain the retreat of the LIS following the LGM. Akçar et al. (2008) also sampled erratic boulders when they constructed a chronology for the retreat of a mountain glacier in Turkey using $^{10}Be$. Late Pleistocene deglaciation in the Wallowa Mountains of Oregon was constrained by Licciardi et al. (2004) using $^{10}Be$ sampled from moraine.

## Problems and Limitations of Cosmogenic Nuclide Dating

A number of problems that affect the accuracy of cosmogenic nuclide ages need to be examined. There is the scatter that is commonly noticed in ages from any given site and this might arise from a number of effects. One is that the inherent assumption in cosmogenic nuclide dating that the surface being dated was exposed continually without interruption is not always valid. In areas where snow cover is experienced during the winter, this assumption may not be accurate. However, if the regional climate is well understood, this effect can be corrected for (Balco, 2011). Another problem occurs when the exposure of a surface being dated was previously interrupted by a cover that no longer exists, such as mobile sediment. An effect of that that type would be difficult to factor into the age calculation. A third problem might be the weathering of the surface by exfoliation, or otherwise, which leads to the partial loss of accumulated nuclides. It was noted in some studies that the scatter in exposure ages from moraine boulders appeared to increase with the age of the surfaces being dated. It is thought that this occurs because the boulders being dated are actually not static throughout the duration of exposure such that the analyzed surfaces have different exposure histories. Hence, in many cases cosmogenic ages do not correspond with the true age of the landforms in question (Balco, 2011).

Another major problem with cosmogenic nuclide ages is the difficulty of determining accurate nuclide production rates at the surface of earth. As discussed above, the absence of a universally accepted scaling model is a problem that remains to be resolved (Balco, 2011). Nonetheless, in all instances, researchers are striving to find solutions to the problems that limit the accuracy of cosmogenic nuclide dating calculations. These include development of statistical methods that allow the scatter in ages to be minimized and continuing to make refinements to scaling models that are used for determining nuclide production rates (e.g., Lifton et al. 2014).

### Future Outlook

Cosmogenic nuclide exposure dating has made a significant impact to glacial geomorphology by making it possible to directly date features that would not have been possible to date previously (Balco, 2011). Nonetheless, problems still remain that will need to be tackled in order to improve the method further. As pointed out above, scaling factors are an essential component of determining exposure ages. However, the lack of a universally accepted model complicates the determination of nuclide production rates. The efforts that are currently being made to address this deficit have the potential to yield ages with improved accuracy in the future, expanding the adoption of the technique by practitioners within the earth sciences. Additional refinements to various other aspects of the technique could also lead to a broader range of geological settings in which cosmogenic nuclide dating could find utility (Ivy-Ochs and Kober, 2008). Lastly, the increasing inventory of glacial chronologies will soon make it possible to model paleoclimate and environmental processes in ways what should bring better insight to late Quaternary processes (e.g., Rupper et al. 2009; Balco, 2011).

## LUMINESCENCE DATING

### Introduction

Glacial sediments that have been mobilized and deposited by either fluvial or eolian processes can be dated using luminescence dating methods. Luminescence dating is a collective term for dating techniques that include thermoluminescence (TL) and optically stimulated luminescence (OSL) approaches. In some literature the terms photon-stimulated luminescence or photoluminescence have been used to describe OSL (Aitken, 1998). Luminescence dating methods make use of the ability of some dielectric materials such as quartz and feldspar to absorb and store energy emanating from environmental ionizing radiation. Such materials are also referred to as dosimeters (Wintle, 2008). In nature, the ionizing radiation comes from the immediate surroundings of mineral grains in their depositional setting as well as from cosmic radiation (Aitken, 1998). Exposure to sunlight leads to the loss of the stored energy such that the minerals have to be buried and remain covered for the energy to start accumulating. If the minerals are collected from their burial site and stimulated using an appropriate mechanism in a laboratory setting, the stored energy can be released as light. Within certain constraints, the light emitted, or luminescence, will be proportional to the energy that was stored (or to the duration of the burial). Hence, if the released energy is measured and the rate at which the energy accumulated is determined, an age representing the time since the energy started accumulating can be calculated. In most cases, this age would be commensurate with the burial or depositional age of the sediment. Hence, in the case of glacial sediment, the age would indicate a time when the sediment was deposited following deglaciation and it would be a minimum age constraint for the retreat of the ice sheet.

The dating range of luminescence dating spans from a few decades up to around 150 ka, subject to certain constraints that are discussed below. However, recent efforts have demonstrated that, in some cases, the upper limit can be extended to a few hundreds of thousands of years (e.g., Murray et al. 2008). Thus, the method is quite effective for dating events of Late Pleistocene or Holocene age. Advantages of luminescence dating of sediments include that the sediments are dated directly, which contrasts with radiocarbon dating. Issues of contamination or transportation which afflict radiocarbon, as outlined in the previous section, are also not a factor. Also, since the method targets inorganic components, datable material can be found in any environment. Recent outlines of the development of luminescence dating and its application in dating sediment have been given by Wintle (2008) and Rhodes (2011).

## Principles of Luminescence Dating

### *The Luminescence Dating Equation*

When subjected to ionizing radiation, minerals such as quartz and feldspar store energy by trapping electrons in lattice defects within their crystal structures. The energy given out by such minerals or dosimeters is referred to as the paleodose. If the rate at which the paleodose accumulated in the mineral under investigation is determined, the luminescence age can be determined using the equation:

$$Luminescence\ age = \frac{Paleodose}{Dose\ rate} \quad (3)$$

For the age obtained to be meaningful, the mineral grain should have been emptied of all previously accumulated energy prior to the event being dated. In nature, as outlined above, this zeroing commonly occurs through exposure to sunlight. Heating to temperatures above 500 °C would also have the same effect. However, with glacial deposits, it is essentially exposure to sunlight that empties mineral grains of previously accumulated energy. Zeroing by sunlight is also referred to as bleaching. Thus, in geomorphological applications, luminescence dating entails determining time that has lapsed since occurrence of a given bleaching event.

### *Electron Trapping Mechanisms in Dosimeters*

Mechanisms involved in the storage of energy through trapping of electrons in lattice defects in mineral grains are not completely understood (Aiken, 1998; Mckeever et al. 1995). Nonetheless, it is thought that ionizing radiation displaces electrons from their parent nuclei into a metastable state. The sites from where the electrons have been evicted serve as holes. The holes and the electrons can diffuse within the crystal separately to ultimately become localized at lattice defects. Types of lattice defects that can lead to such trapping include a negative ion that is missing from its appropriate site in the lattice, a negative ion located in an interstitial site, or impurity atoms that occur in the lattice as a result of substitution (McKeever et al. 1995). Other trap types have been reported (e.g., McKeever et al. 1995) but these are beyond the scope of this chapter. An energy level diagram used to depict the creation of luminescence centers in crystal lattices is shown in Figure 4. The efficacy of a trap, which is measured by 'E,' corresponds to the depth of a trap (T) beneath the conduction band. Traps are termed stable if they are able to withstand the effects of disruptions that could evict electrons from the traps. Stimulating the material using a suitable energy source such as by heating to an adequately high temperature or by using an optical source of appropriate wavelength, can result in the electrons being dislodged from their traps. Evicted electrons diffuse within the crystal lattice until they arrive at a site with an affinity for electrons. Such sites are termed recombination centers. When some recombination centers merge with electrons, the process is accompanied by the release of light. Localized holes usually form effective recombination centers for luminescence dating purposes. The diffusion occurs so rapidly that the interval between stimulation and recombination can be deemed immediate.

An important aspect that has methodological connotations and has been alluded to above is that stimulation of energy release to obtain a signal during luminescence measurement can be conducted by either heating, which gives TL, or by optical means which gives OSL. The intensity of the signal that is obtained following stimulation is proportional to the number of electrons that are stored in the mineral grain being analyzed, which in turn corresponds to the duration of exposure to ionizing radiation, or burial age of a material.

An important aspect of the energy storage mechanism is that the number of traps within any dosimeter is finite. Hence, accumulation of energy cannot continue indefinitely. At some point, the traps become exhausted and this saturation essentially determines the upper age limit that can be attained using luminescence dating methods.

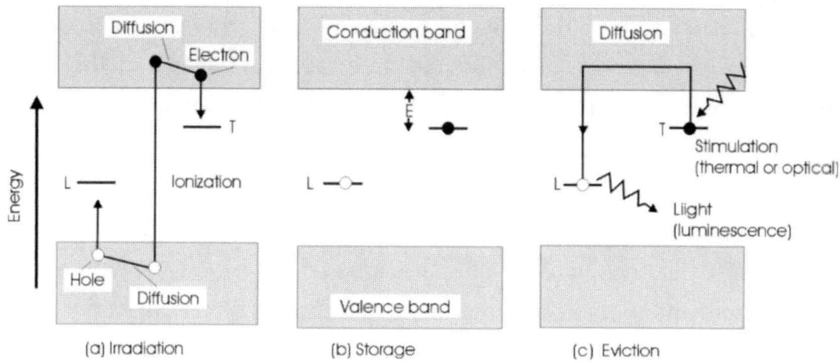


Figure 4. Energy Level diagram illustrating the creation of luminescence centers in crystal lattices through exposure to ionizing radiation (redrawn from Aitken (1998)). (a) Irradiation leads to electrons being expelled from their original site to diffuse within the lattice. (b) Electrons become trapped while holes of missing electrons become localised at particular centers. (c) Stimulation (by heat or light) results in electrons being evicted from the traps and diffusing until they meet a recombination center. Light (luminescence) is emitted when luminescence centers are encountered by the electrons to give TL or OSL.

## Sources of Ionizing Radiation in Nature

In natural settings, ionizing radiation that contributes to the paleodose received by mineral grains includes isotopes of the uranium ($^{238}U$ and $^{235}U$) and thorium ($^{232}Th$) decay chains, potassium ($^{40}K$) and rubidium ($^{87}Rb$). These elements typically occur in earth materials in very low concentrations. However, the radiation that they give out, which includes alpha and beta particles as well as gamma radiation, is adequate to yield signals that can easily be measured using appropriate equipment. A smaller additional component of radiation received by earth materials comes from cosmic rays.

## Luminescence Measurement Equipment

In order to determine the age of a given sample, sediment collected from the field is first pretreated as will be discussed below. A specialized luminescence reader is then used to measure the energy stored in the mineral grains. As outlined above, there are two main methods for stimulating energy release from the mineral grains when conducting measurement. Heating the sample enables TL to be acquired. Stimulating the source with an optical source such as blue light or infrared radiation allows OSL to be released and measured. Figure 5 presents the basic layout of a luminescence reader. Standard instruments are usually equipped with both thermal and optical stimulating capabilities. Prepared samples are loaded onto small discs, about 1 cm in diameter, that are then placed onto an appropriate holder before being introduced into the reader. Within the instrument, the sample discs, which are usually introduced in multiples, are selectively moved into position for measurement using an automated system such that a large number of discs can be analyzed in a relatively short period of time. Following stimulation, the light or luminescence signal emitted by the sample is detected using a system such as a photo multiplier tube (PMT) or a charge couple-device (CCD) camera (e.g., Richter et al. 2011). Filters are used to focus on parts of the electromagnetic emission spectrum that are of interest.

Sample stimulation using TL entails heating the sample at temperatures of up to 20 °C per second. When a temperature that corresponds with the trap depth ('E' in Figure 4) is reached, the trapped electrons are quickly evicted into the conduction band and may ultimately reach a recombination site where luminescence occurs. On a curve showing temperature versus the TL signal (in photon counts), the eviction of electrons will appear as an emission peak. Such a plot is referred to as a glow curve. If the sample is heated again just after the first heating, it will yield no luminescence peak because the first heating would have emptied any stored dose, or zeroed it.

When stimulating samples using optical means, a source of a preferred wavelength is chosen. Blue, green and infra-red wavelength sources are commonly employed. The sensitivity of a trap to optical stimulation is usually dependent on the type of the trap as well as on the wavelength of the source. Generally it is observed that electrons are evicted more rapidly when shorter wavelengths are used. Optical stimulation sources are also usually chosen to allow separation to be made between wavelengths of the stimulation source from that of the signal of the mineral being analyzed. Optical filters may also be used to enhance the separation.

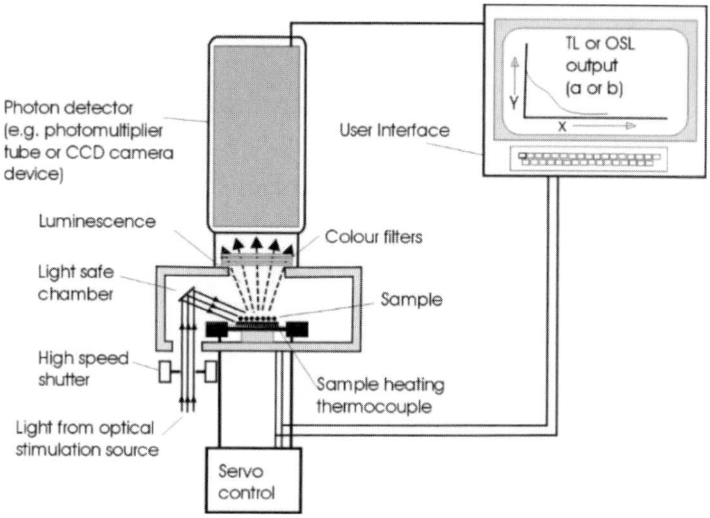


Figure 5. Basic components of a luminescence reader system. Note that the reader can employ either thermal or optical stimulation.

## Advantages of OSL Compared to TL

When dating sediments that have been reset or zeroed by sunlight, OSL dating has a number of distinct advantages compared to TL. First, solar bleaching of the luminescence signal that is sensitive to optical stimulation occurs more rapidly than the signal that is stimulated by heating (Godfrey-Smith et al. 1988). This is because optically sensitive electron traps are not necessarily the same as those that are sensitive to thermal stimulation. Hence, chances for samples that are analyzed using optical means to be well-zeroed by exposure to sunlight in nature are greater than for those that are dated using thermal means. Second, thermally stimulated signals have a higher residual signal following analysis which could make it difficult to date very young samples using TL

(Aitken, 1998). Thus, the preferred measurement method for samples that have not previously been zeroed by heating is optical stimulation. When analyzing samples that were zeroed by heating however, such as when dealing with fired pottery or baked sediments, TL is preferred.

## What Minerals Can Be Analyzed Using Luminescence Methods?

There are many minerals that will luminesce when stimulated with a suitable source following exposure to ionizing radiation. Nonetheless, not all of them can be used for luminescence dating. To date, only quartz and feldspar dating have been developed for routine chronological studies. There have been efforts to use zircon and calcite but there are still a number of technical difficulties that need to be resolved before the minerals can be used for day-to-day analysis.

Quartz is by far the most widely used mineral in luminescence dating studies because it possesses a number of positive attributes compared to other minerals. First, quartz is very stable at the earth's surface, being resistant to both physical and chemical weathering. Consequently, it occurs in abundance in most clastic sediments. Second, quartz possesses luminescence properties that are very stable. Third, quartz does not usually have an internal source of radiation which simplifies dose rate calculations. However, there are some instances where some low concentrations of uranium may be present but these are rare (Aitken, 1985).

It is important to note that the OSL signal of quartz has been reported to consist of at least three components that are described as fast, medium and slow with reference to their decay rates (Bailey et al. 1997; Jain et al. 2003). Hence, following exposure to sunlight, the fast component would be the first to be depleted.

Feldspar is also widely used in luminescence dating. The mineral makes up about 60% of the earth's crust and, though it is more susceptible to both chemical and physical weathering than quartz, it still makes up a significant proportion of minerals found in clastic sediments. Compositionally, feldspar has a variety that contains potassium (K).

Because potassium has an unstable isotope that produces ionizing radiation, dose rate calculations have to account for that internal potassium when dealing with samples containing K-feldspars. As a solution, when dating feldspars, K-feldspars have to be separated from Ca- and Na-feldspars and analyzed separately. In comparison with quartz, there are some attractive characteristics that feldspar possesses. First, feldspar luminescence signals are usually brighter than those of quartz. Consequently, smaller doses can be measured. Second, feldspar responds to stimulation using the near-infrared part of the electromagnetic spectrum. This facilitates the separation between stimulation source wavelengths and emissions wavelengths which are shorter. Third, the internal dose from $^{40}K$ in K-feldspars allows part of the dose rate to be estimated more accurately as this is not affected by external effects such as pore water. The main disadvantage of working with feldspar however, is that it experiences a phenomenon called anomalous fading (Wintle, 1973). Through anomalous fading, the stored dose leaks such that, following irradiation, the paleodose decreases over time. There are corrections that can be used to account for anomalous fading (Lamothe and Auclair, 1999).

## Determination of Paleodose and Dose Rate

As outlined in Equation 3, the luminescence age is calculated by dividing the paleodose by the dose rate experienced by the mineral grains. This section examines methods that are used to determine the paleodose and the dose rate.

### *Paleodose Determination*

For a luminescence age to have some significance, the accumulation of the paleodose should coincide with a geomorphic event that zeroed the grains of any energy that might have been acquired previously. The dose is measured using units of Gray (Gy) where 1 Gy = 1 J/kg and it refers to energy acquired per kilogram of material. When determining the paleodose of a sample of unknown age, a sample aliquot is measured in a laboratory setting first to determine the level of luminescence intensity that it gives out. This signal emanating from the natural dose in the field is referred to as the natural signal. Aliquots of the sample are then taken and given well calibrated artificial doses in the laboratory and also measured. By comparing the artificial dose signal levels with the signal from the natural sample, an artificial dose level that would give out a signal equal to that obtained from the natural sample is determined. This dose level that corresponds to the natural signal is referred to as the equivalent dose ($D_e$). There are two main approaches that are employed for determining $D_e$ and these are the additive dose and the regenerative (or regeneration) dose methods (Aitken, 1998).

### *Additive Dose Method*

To determine $D_e$ using the additive dose method, samples are split into multiple aliquots after which a few are taken and measured using a luminescence reader to determine the natural signal. The remaining aliquots are given incremental artificial doses on top of the natural signal using an irradiator. These aliquots are then also measured. The signals obtained are plotted against the artificial dose that was given. The resulting plot is known as the growth curve (Figure 6). The natural signal is plotted against the zero laboratory dose on the growth curve and if the curve is extrapolated backwards, it intercepts the horizontal axis at a dose level equal to $D_e$. Since the sensitivity of mineral grains varies between samples, each sample is treated as unique and a new growth curve is generally required with each age determination.

### *Regenerative Dose Method*

The regenerative method follows an approach similar to the additive dose method in that the sample is split into multiple aliquots after which the natural signal of a few aliquots is measured. Where it differs from the additive dose method is that before administering the artificial dose on the remaining aliquots, any natural signal that the aliquots may have is first zeroed. When dealing with materials that were previously reset by sunlight, exposure to sunlight is usually employed for zeroing. Following measurement, the acquired artificial dose signals are also plotted to give a regenerative dose growth curve. The equivalent dose is determined by interpolating the natural signal into the growth curve as shown in Figure 6b.

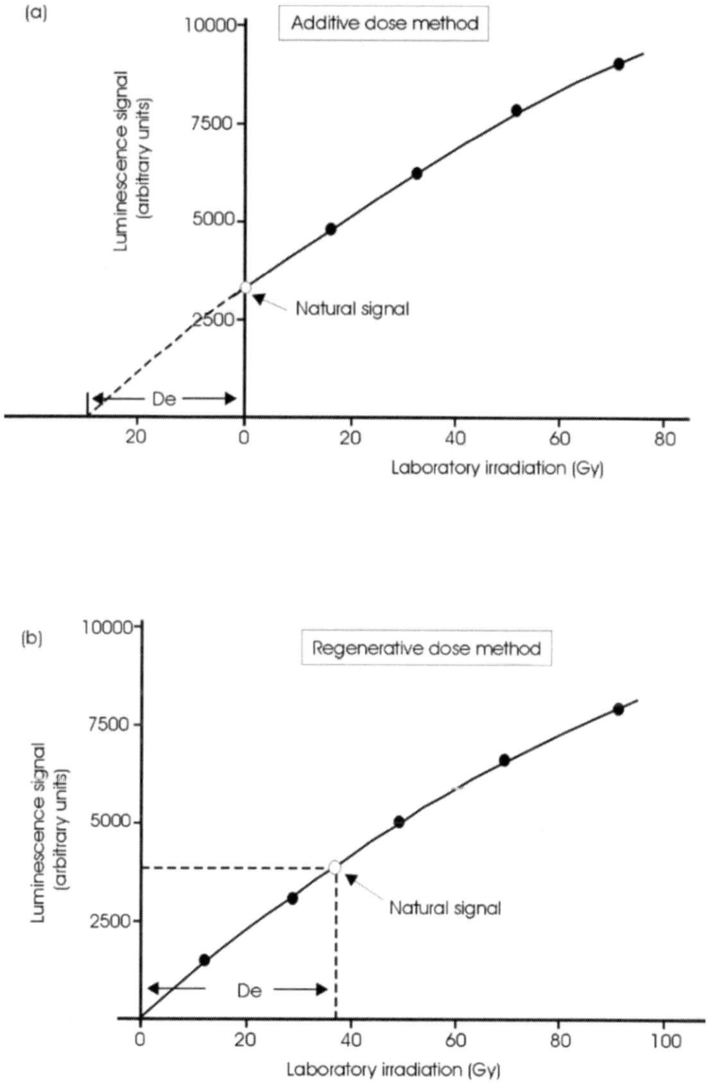


Figure 6. Determining the paleodose (De) using (a) the additive dose method and (b) the regenerative dose method.

***Single Aliquot and Single Grain Dating Methods***

Luminescence dating began with TL dating of archeological artifacts (e.g., Aitken, 1985) after which OSL dating was subsequently introduced. When initially developed for TL dating, the additive dose and regenerative dose methods outlined above employed multiple aliquots for each dose level after which mean values were calculated. However, when OSL dating emerged (Huntley et al. 1985), it was noted that it was possibile to employ a single aliquot for constructing the growth curve. The use of a single aliquot eliminates the inter-aliquot variations that are observed when working with multiple aliquots which increases uncertainties. However, the use of single aliquots when working with the regenerative dose was initially hampered by the change in sensitivity of an aliquot following multiple cycles of measurement and irradiation. The problem was ultimately resolved by monitoring the sensitivity change by systematically administering a test dose to the aliquot that was then used for correction (Murray and Roberts, 1998). Following some refinements, the single aliquot regenerative (SAR) protocol (Murrary and Wintle, 2000, 2003) was developed and it has since emerged as the preferred method for routine analysis in luminescence dating.

The SAR protocol has also been employed for determining the paleodose for individual sediment grains as opposed to using multigrain aliquots. The approach is usually expedient when working with a sediment whose grains have different paleodoses (Olley et al. 2006). This is often witnessed in materials that are partially bleached as is often observed with fluvial deposits.

## Determining the Dose Rate

As outlined above, the energy that accumulates in luminescence dating dosimeters emanates from ionizing radiation from the immediate surroundings of the mineral grains. The ionizing radiation includes energy from the radioactive decay of isotopes in the uranium ($^{238}U$ and $^{235}U$) and thorium decay chains, as well as the decay of $^{40}K$ and $^{87}Rb$. A number of methods can be used to determine the contributions of these isotopes. These include the concentration approach, which measures the concentrations of the radioactive elements in the sediment using various analytical chemistry techniques such as atomic absorption spectroscopy (AAS), neutron activation analysis (NAA) and flame photometric detection (FPD) (Aitken 1998). However, such methods have limitations especially for detecting uranium due to disequilibrium (Aitken, 1998).

Another approach for determining the dose rate is to directly measure the activity of the radionuclides using alpha or gamma ray spectroscopy. Additional methods for determining the dose rate exist, including some that involve taking measurement in the field, which is the preferred approach whenever possible. Additional details on dose rate measurement methods can be found in Fuchs and Owens (2008).

## Sample Collection and Preparation for Analysis

### *Sample Grain Size*

As outlined above, luminescence dating for geological applications provides ages that are commensurate with an energy zeroing event that occurred in the past. Materials that are well zeroed are usually those that comprise grains of sand or smaller grain size. Thus, there are generally two main grain size categories that are employed in dating. These are termed the coarse grain and the fine grain approach. The coarse grain approach employs sand sized grains in the diameter range 90-250 μm. When using coarse grains, the outer rim of the grains that is affected by alpha radiation is removed by etching such that only the inner parts of the grain that received the beta, gamma and cosmic radiation is analyzed. For that reason, coarse grain dating is sometimes referred to as inclusion dating (Fleming, 1970). The age equation used for coarse grain dating is modified to:

$$Luminescence\ age = \frac{Paleodose}{0.9D_{\beta}+D_{\gamma}+D_{C}} \quad (4)$$

where $D_\beta$, $D_\gamma$ and $D_C$ are the contributions to the dose rate from beta, gamma and cosmic radiation. The beta contribution is factored by 0.9 to account for some of the beta irradiated part of the sample eliminated by the etching that removes the outer rim of the grains penetrated by alpha particles.

Typically, coarse grain dating uses pure mineral separates because quartz and feldspar have different luminescence properties. Fine grain dating uses silt size sediment particles in the 4-11µm range. In this case it is not necessary to etch the outer rim of the particles because alpha particles penetrate the grains entirely. As a result, the age equations for fine grains is:

$$Luminescence\ age = \frac{Paleodose}{kD_\alpha + D_\beta + D_\gamma + D_C} \tag{5}$$

where $D_\alpha$, $D_\beta$, $D_\gamma$ and $D_C$ are contributions to the dose rate from alpha, beta, gamma and cosmic radiation to the dose rate. There are some studies that have used grains in the silt size range 33-63 µm which has been referred to as middle-grain dating (Aitken, 1998). In all cases, interstitial water in the sediment attenuates the dose received by the mineral grains. Hence, field water content in the sediment has to be determined and factored into the dose rate calculation.

### *Sample Collection and Pretreatment*

The most important requirement when collecting field samples for luminescence dating is that samples must not be exposed to sunlight, otherwise the stored energy will be lost. This places constraints on the sampling procedures and a number of techniques have been devised to fulfill that requirement. One approach is to collect samples at night. However, this is particularly inconvenient and increases the chance of accidental exposure to light. Thus, nighttime collection is no longer widely practiced and most practitioners collect samples during the day. When working with sediment that is relatively firm, one can cut out a block of sediment and take it to the laboratory. There, a sample for analysis would then be extracted from the interior of the block that was not exposed to light. A technique that is now widely used for collecting samples for luminescence dating from sediment is to insert an opaque metallic or plastic pipe into the sediment and then cap the ends with a light-free seal after extracting it. At the lab, the sediment for dating is extracted from the interior of the pipe for processing. Where access to targeted depositional units is limited, samples for dating have also been collected by drilling vertically using specially designed modules for extracting samples at depth (e.g., Munyikwa et al. 2011b).

Before luminescence analysis can be performed on the collected samples, they have to be processed under safe-light conditions. As described earlier, coarse grain samples are usually analyzed as pure mineral separates of either quartz, K-feldspar or Ca- or Na-feldspar. Hence, the pretreatment entails removal of carbonates and organic material using dilute hydrochloric acid (HCl) and hydrogen peroxide ($H_2O_2$) respectively. Sieving is used to separate grain sizes in the desired range and a heavy liquid such as sodium polytungstate solution is then used to separate pure quartz or feldspar. For analysis, the separated grains are mounted on discs. Further details on sample preparation can be found in Stokes (1992).

When working with fine grains, carbonates and organic material are also removed using dilute HCl and $H_2O_2$ respectively. Dilute sodium oxalate may be used to avoid flocculation of the particles. Grains in the size range 4-11 μm are then extracted using a sedimentation column and mounted on discs for analysis. In some studies, pure quartz has been extracted from silt size sediment by digesting feldspar using flourosilicic acid (e.g., Rees-Jones, 1995).

## What Materials Can Be Dated Using Luminescence Methods?

A geological material has to fulfill a number of criteria before it can successfully be dated using luminescence techniques. First, the material should contain mineral grains that can store energy from ionizing radiation. As pointed out above, quartz and feldspar usually satisfy this requirement when working with clastic sediments. Second, there must have been a zeroing event in the grain's depositional history during which all previously acquired energy was emptied. This would normally be an event of geomorphic significance that exposed the grains to sunlight or heated the grains to temperatures exceeding around 500 °C. The third and final requirement is that the electron traps in the mineral grains should not have been exhausted by the time the dating is performed. In essence, the storage of energy should not have reached saturation.

Given the above criteria, luminescence dating materials in earth sciences generally fall into two main categories: those that were zeroed by heating and those that were reset by exposure to sunlight. Heated earth materials include sediments that experienced elevated temperatures at some point within the last ca. 100-200 ka, for instance by volcanic eruptions. Products of volcanic eruptions such as lava or ash are also included in this category. Materials affected by meteorite impacts that result in elevated temperature would also fall into this group. It has to be mentioned that luminescence dating has also been applied extensively in archaeology where heated materials such as ancient pottery, fired bricks, tiles and other heated artifacts provide very useful chronologies.

The second class of earth materials that can be dated using luminescence methods comprises sediments that are reset by solar bleaching. These are usually sediments that are exposed to sunlight during transport and deposition. Once buried by overlying sediment, the geological clock starts counting through the accumulation of energy from ionizing radiation. Deposits that fall into this category include sediments transported by wind (eolian sediments), sediments transported by water, sediments associated with glaciers and earthquake related deposits. Since the focus of this chapter is on dating deglaciation, the discussion below will only examine dating of sediments that may be associated with glaciers.

## Constraining the Last Deglaciation Using Luminescence Dating

As outlined in the introduction, constraining deglaciation can be achieved by dating sediments that were directly deposited by the glacier as well as those sediments that were emplaced by glacial meltwaters (Thrasher et al. 2009). There have also been efforts made to constrain deglaciation by analyzing sediments that were emplaced by glacial processes on a postglacial surface after which they were reworked by eolian activity (e.g., Munyikwa et al. 2011a; 2017). In all instances, as will be outlined below, the postglacial sediments provide minimum chronologies by which glaciers or ice sheets had receded from the landscape.

### *Dating Sediment of Glacial Origin*

Glacial deposits can comprise a wide range of particle sizes ranging from boulders and gravels to sands, silt and clay. Such clastic deposits commonly contain quartz and feldspar, which would in theory make it possible to analyze them using luminescence methods. However, focus would need to be placed on the finer fractions as these provide the best chance for adequate bleaching. Sediments deposited by glaciers and ice sheets can be transported on top of the ice, within the ice or underneath the ice (Thrasher et al. 2009). Sediment transported underneath the ice or subglacially, includes deformation till as well as meltout till. Both subglacial and meltout till are usually emplaced as thick sequences that experience minimal exposure to sunlight (Thrasher et al. 2009). As a result, the poorly bleached sediments would not be ideal for dating using luminescence methods (Tsukamoto et al. 2002; Zhou et al. 2008). For instance, deformation tills in eastern Nepal were dated by Tsukamoto et al. (2002) using OSL and found to yield chronologies that overestimated their true age significantly. Water-lain deposits that are deposited under glaciers such as eskers and fluvio-kames have also been found to be very poorly bleached and not suitable for luminescence dating (Raukas and Stankwoski, 2005).

Sediment that is transported on top of the glacier has a better chance of being zeroed by sunlight. Such sediment could be spread across the surface of the glacier or ice sheet by supraglacial stream flow (Thrasher et al. 2009; Zeng et al. 2016). However, if additional snow falls on the glacier surface, burying the supraglacial sediment, adequate bleaching of the sediment might not occur. Sediment could also be placed on top of a valley glacier by slumping of valley walls. Nonetheless, exposure to sunlight for slumped sediment would most likely be limited.

Studies have demonstrated that glacigenic sediments that have the highest chance of being well-zeroed are sediments that are transported away from the glacier by meltwaters as outwash (Richards, 2000; Fuchs and Owen, 2008; Thrasher et al. 2009; Zeng et al. 2016). The likelihood that glaciofluvial outwash will be well-bleached increases with distance away from the glacier's front (Rhodes and Bailey, 1997; Thrasher et al. 2009; Munyikwa et al. 2011b). Proglacial outwash sediments can be transported for several kilometers from the ice sheet or glacier margin which increases the chance for exposure to sunlight. Consequently, most studies that have attempted to date deposits of glacial origin have been directed at this class of sediments (e.g., Duller, 2006; Klasen et al. 2006; Alexanderson and Murray, 2007; Bøe et al. 2007; Pawley et al. 2008; Zhang et al. 2012; Gaar et al. 2013; King et al. 2013; Bickel et al. 2015; Zheng et al. 2016). Turbulence during high flow conditions in glaciofluvial systems, however, could also mean that some grains are not adequately exposed to sunlight, though some near the top might be well bleached (e.g., Sanderson et al. 2007; Raukas et al. 2010). Overall, however, it is the consensus that adequate bleaching can occur in outwash systems that would allow luminescence dating methods to be applied. Finer grains and environments with lower turbidity would have conditions where zeroing would be more likely to occur (Rhodes and Bailey, 1997). Sandurs are outwash plains that are formed by waters released by melting ice sheets. Under low flow conditions, some parts of a sandur may become subaerially exposed and, should wind velocities suffice, some grains could be mobilized by wind, increasing their chances of being zeroed (Richards, 2000; Thrasher et al. 2009). An effect that might aid bleaching of glaciofluvial deposits is also that most glacier melting has been noted (e.g., Richards, 2000) to occur in the summer and during the warmest part of the day when, in both cases, insolation is most intense.

When dating glaciofluvial deposits that have not been completely zeroed, it has been noted that employing single-grain dating methods helps identify grains that have been well zeroed (e.g., Duller 2006). Another problem that has been observed in some studies of glaciofluvial deposits is the low sensitivity of the quartz from some regions (e.g., Rhodes, 2000; Richards, 2000; Spencer and Owen, 2004; Lukas et al. 2007). It has been suggested that the low sensitivity occurs in quartz that is derived from geologically young sources that are non-sedimentary, such as metamorphic bedrock. Others have proposed that the low sensitivity is found in grains that have not experienced multiple bleaching and dosing cycles prior to their burial (e.g., Pietsch et al. 2008). Glacigenic quartz in some studies has also been shown to lack or to display a very weak fast component (e.g., Lukas et al. 2007). Hence, Thrasher et al. (2009) think that the sensitivity of quartz is influenced by two main factors. First, quartz from some provenances has inherently poor sensitivity for luminescence dating purposes and second, an increase in the number of erosional and depositional cycles that the sediment is subjected to prior to burial increases the sensitivity of quartz.

### *Dating Eolian Deposits on Postglacial Landscapes*

As discussed earlier, glacial sediments that have been reworked and redeposited by wind in the immediate aftermath of the retreat of the glacier offer a convenient proxy for constraining glacial retreat by providing minimum ages for ice sheet recession. Eolian deposits are the sediment of choice for dating using luminescence methods and these can be in the form of loess or dune sands (Aitken, 1998). This is because both types of sediment are transported subaerially and the exposure to sunlight that they experience during the transport usually afford them adequate time for bleaching (e.g., Singhvi and Porat, 2008). For purposes of constraining glacial retreat, focus in this discussion will be placed on eolian dune deposits as it is easier to link source area of the sediment to formerly glaciated landscapes.

The retreat of glaciers leaves behind large areas covered by till and glaciofluvial sediments. Typically, such areas would also initially be devoid of vegetation. The drainage of proglacial lakes can also leave behind large areas blanketed by deltaic sands and other fine grained glaciolacustrine sediments. Prior to the colonization of the landscape by vegetation, strong winds on the periphery of the retreating ice sheet could mobilize such exposed sediments and deposit them in dune bedforms (Cooke, 1993). Main requirements for the mobilization of sediment by wind include the availability of grains of appropriate size, low moisture content (Belly, 1964; Azizov, 1977; Logie, 1982) and low vegetation cover. As will be discussed later, this was particularly the case on the western margin of the LIS following the termination of the LGM. Hence, it is plausible to assume that eolian processes would have prevailed during the period that immediately followed the retreat of the ice sheet. By inference, the occurrence of fossil dune landforms in areas that were previously covered by ice sheets points to ice-free conditions at the start of dune formation. If luminescence dating can be used to provide chronologies for the dune formation, the ages can serve as minimum age constraints for the retreat of the ice sheets.

A problem with using eolian dune chronologies for dating deglaciation is that dune deposits are often reworked by subsequent eolian phases. Thus, the eolian deposition does not necessarily coincide with deglaciation. Rather, the ages from the eolian deposits simply provide minimum constraints by which the ice sheet had receded from the landscape.

Investigations in which luminescence ages have been used to constrain deglaciation include studies by Munyikwa et al. (2011a, 2017) where eolian dune depositional chronologies in western Canada have been used to constrain the retreat of the LIS following the LGM. A case study is presented later in this chapter looking at the study by Munyikwa et al. (2011a, 2017). On the British Isles, Telfer et al. (2009) dated loess deposits from NW England that is thought to have been deposited during deglaciation of the region following the last ice age.

## Limitations for Luminescence Dating and Future Outlook

Luminescence dating has made immense contributions to the study of paleoenvironments over the last several decades by providing a robust chronometer in situations where no other dating method could have produced any functional temporal constraints. The chronologies have enabled quantitative studies to be undertaken in ways that were not previously possible. Contributions to glacial processes have also been significant (Fuchs and Owen, 2008; Thrasher et al. 2009). This, however, is not to say that the dating method is not without complications, especially when it comes to dating features related to glaciation. As outlined above, partial bleaching is a problem that is particularly acute when dealing with glacial deposits. This is because glacial sediments are often not exposed to adequate sunlight. Even for reworked glacial sediments such as outwash deposits, glaciofluvial sands often are not completely bleached such that researchers have often resorted to using single grain dating approaches that try to identify individual grains that were well-zeroed (e.g., Fuchs and Owen, 2008; Thrasher et al. 2009).

A general characteristic of luminescence dosimeters that can be a limitation is that, as discussed above, once traps in mineral grains become exhausted, the storage of energy in the grains ceases to be reliably proportional to the period of exposure to ionizing radiation (Aitken, 1985). This saturation places a cap on the upper age limit of events that can be dated using luminescence techniques. Quartz for instance will saturate at around 100-150 Gy, though this may vary depending on the type of quartz in question or the provenance. Feldspars may saturate at higher doses, but that too places a limit beyond which one can produce a meaningful age.

While this would not be an impediment for dating features from the last deglaciation, unless the dose rates are exceptionally high, obtaining ages for glacial events older than ca. 100 ka, for instance, could be less practical.

It has to be mentioned that there are ongoing efforts to gain a better insight into the behaviour of both quartz and feldspar at high doses (e.g., Murray et al. 2008; Timar-Gabor et al. 2015; Buylaert et al. 2011; 2015). Such studies should lead to improvements in the accuracy of older luminescence ages. An enhanced understanding of the behavior of both quartz and feldspar should also help produce more accurate luminescence ages in general. There are recent studies (Frouin et al. 2017) that have used the infrared radiofluorescence signal of K-feldspars for dating as an alternative to the more established IR signal. The infrared radiofluorescence signal is thought be less susceptible to anomalous fading and, if the technique were to be refined for routine use, it would significantly broaden opportunities in luminescence dating (Frouin et al. 2017).

# Radiocarbon Dating

## Introduction

The relative simplicity of radiocarbon dating compared to the other two methods described so far has resulted in it being the most widely utilized absolute age dating method in constraining deglaciation. Its widespread use is also because it was one of the earliest radiometric dating methods to be developed for routine use. The technique was discovered through work conducted by Willard Libby between 1946-1949, at the University of Chicago, who was examining the prevalence of $^{14}C$ in nature (Taylor, 2016). Earlier work in the 1930s had demonstrated that $^{14}C$ could be created in particle accelerators but it was not known if the isotope was present in the natural environment. Hence, in 1946, Libby had commented that if the half-life were significantly higher than 1 ka, an equilibrium would exist between the production and decay of $^{14}C$ in living organisms (Libby, 1946). Libby also noted that upon death of an organism, the activity of $^{14}C$ would diverge significantly from that of live ones (Taylor, 2016). Between 1946 and 1948, work conducted by Libby's research group demonstrated that methane from modern sewage from Baltimore, Maryland had significant activity from $^{14}C$ while methane from petroleum showed no $^{14}C$ activity. This was an indication that $^{14}C$ in dead organisms decayed with time.

Another critical experiment conducted by Libby's group was to check whether the $^{14}C$ activity that is found in living organisms varied with geomagnetic latitude since the production of radiocarbon in the atmosphere was not constant, with the highest rate being at the poles and lowest at the equator. The test was conducted by measuring the specific activity of $^{14}C$ in disintegrations per minute per gram of carbon on organic samples from various sites around the world. The results showed that levels of $^{14}C$ in living organisms were globally homogenous, irrespective of geographic location. This suggested that once produced in the atmosphere, $^{14}C$ would defuse rapidly to reach a global uniformity close to the earth's surface.

The final step of major significance for the development of radiocarbon dating performed by Libby's group was to try and date samples of known age using the fledgling dating method. Samples analyzed included wood from the Egyptian tomb of Pharaoh Djoser thought to date from around 2.6 ka BC. The results were reported to be congruent with the expected age (Libby et al. 1949) which served as confirmation that the scientific basis of the approach was sound. Subsequently, radiocarbon age determination was used on additional samples of known age (Arnold and Libby, 1949; Libby, 1952) with positive results such that between 1949 and 1954, the method was being used to date samples of unknown age. After 1954, radiocarbon dating gained global acceptance and, with further refinements, the method has evolved into the most well-known and widely used radiometric dating technique (Walker, 2005).

## Basic Principles of Radiocarbon Dating

Natural carbon comprises three main isotopes: $^{12}C$, $^{13}C$ and $^{14}C$. In terms of abundance, $^{12}C$ makes up around 98.89% while $^{13}C$ constitutes about 1.11%. Only a very small fraction (0.0000000001%) is from $^{14}C$. Of the three, isotopes, $^{12}C$ and $^{13}C$ are stable and the third, $^{14}C$ (radiocarbon), is unstable. There is no significant difference in the chemical properties of the three isotopes.

Radiocarbon is formed in the upper atmosphere when neutrons from cosmic rays interact with nitrogen atoms ($^{14}N$). The reaction is represented by the equation:

$$^{14}_{7}N + n \rightarrow \, ^{14}_{6}C + p \tag{6}$$

where $n$ is a neutron from cosmic rays and $p$ is a proton. The $^{14}C$ that is produced rapidly reacts with oxygen that is in the atmosphere to form carbon dioxide ($^{14}CO_2$) which diffuses throughout the atmosphere and mixes with other $^{12}CO_2$ and $^{13}CO_2$. The dispersal process is very rapid, as was verified by measurements of radiocarbon released in the testing of thermonuclear devices in the 1960s (Nydal, 1966). Plants assimilate the $^{14}CO_2$ through regular photosynthetic pathways after which it enters other food chains. The earth's oceans assimilate a very large proportion of the $^{14}C$ (>95%) through atmospheric exchange processes and as dissolved carbonate. This means that marine organisms such as corals and molluscs, foraminifera, etc. also take up $^{14}C$ during their lifetime. In essence, $^{14}C$ joins the carbon cycle where it is exchanged within the biosphere, geosphere and atmosphere. Living organisms, both faunal and floral, that assimilate carbon through photosynthetic processes, or when plant-eating animals consume food, constantly take up $^{14}C$ during their lifetimes. Animals that consume other animals, for instance carnivores, also take in $^{14}C$ through the consumption of plant-eating organisms. As a result of the connections in the food chain, there is an equilibrium between the concentration of $^{14}C$ in the atmosphere and that which is in all living organisms. Thus, the abundances of $^{12}C$, $^{13}C$ and $^{14}C$ in the atmosphere and in living plant and animal tissue is approximately the same. When an organisms dies, however, the active uptake of carbon by the plant or animal ceases such that the decaying $^{14}C$ is no longer replenished to maintain environmental levels.

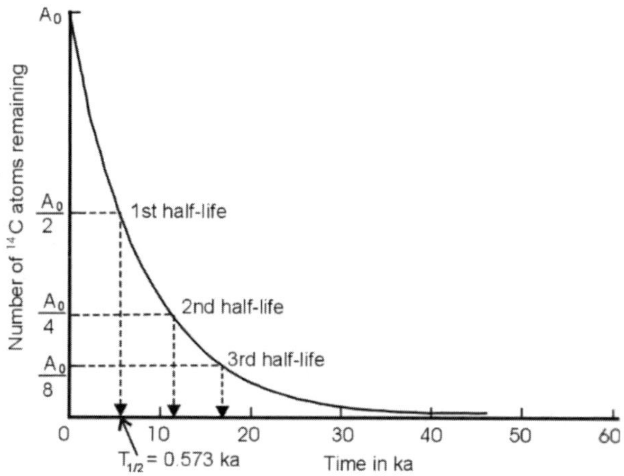


Figure 7. The decay curve for radiocarbon. Because the decay is constant, the percentage decrease of atoms within a given period does not change. Hence, the curve has an exponential form.

Radiocarbon decays by emitting beta particles to yield nitrogen according to the equation:

$$^{14}_{6}C \rightarrow\ ^{14}_{7}N + e^{-} + \bar{\nu}_e \qquad (7)$$

where $e^-$ is an electron (or beta particle) and $\bar{\nu}_e$ is an electron antineutrino.

The beta particles released by the process are relatively weak, typically possessing a maximum energy of around 160 keV. With such an energy profile, the beta particles can travel a maximum distance of 22 cm in air and only 0.22 mm in body tissue. The decay of radiocarbon occurs at a constant rate. Libby et al. (1949) originally measured the half-life of $^{14}C$ to be 5568 ± 30 years. However, more recent measurements have shown that the half-life is around 5730 years. The constant decay rate yields an exponential decay curve (Figure 7). For dating purposes, an important consequence arising from this half-life is that after about 10 half-lives, only a very small amount of radiocarbon remains in the sample to the extent that it would be impossible to measure. This corresponds to about 50 ka (Figure 7) and places an upper age limit of about 45 ka on materials that can be dated using radiocarbon. Consequently, the method can only be used for dating materials from the later half of the Late Pleistocene. Older materials would require the use of an alternative dating method.

## Measuring Radiocarbon

Dating an object using radiocarbon requires determining the amount of $^{14}C$ that still remains in the object and comparing this level with the concentration at the time the organism died or became isolated from the global equilibrium. Such materials can include wood, charcoal, peat, plant remains, organic lake sediment, shell, coral, etc. There are two main methods that are used for determining the amount of $^{14}C$ that still remains in a material that is being dated (e.g., Theodórrson, 1991). One approach is to measure the activity of $^{14}C$ that still remains in the material by counting the beta particles being released within a given period of time. The rationale of this approach is that the rate with which the beta particles are being emitted reflects the level of activity and, hence, the concentration of $^{14}C$ that still remains in the material.

The second approach is to use AMS which measures the relative proportions of carbon isotopes in a given sample. Thus, in this instance, the concentration of $^{14}C$ remaining in the sample is obtained directly whereas in the beta counting procedure, the concentration of residual $^{14}C$ is determined from the activity of radiocarbon.

### *Beta Counting*

When radiocarbon dating was first developed, beta counting was employed for determining the residual activity of $^{14}C$ in dating samples because natural levels of radiocarbon were too low to measure using conventional mass spectrometers. With isotopic ratios of less than $10^{-10}$, $^{14}C$ would be too low for the detection limits of regular mass spectrometers. There are two main ways of conducting beta counting (Walker, 2005). One is to use gas proportional counting whereby the sample is first converted into a gas such as carbon dioxide, ethylene, methane or ethane which is then introduced into a counting chamber. In the counting chamber, beta emissions from the decaying $^{14}C$ are detected by a charged wire that is centrally positioned. It may be of interest to note that when gas proportional counting was introduced in radiocarbon dating in the 1950s, it was replacing solid carbon counting that had been used by Libby and his group (Theordórsson, 1991).

The second beta counting method, which has largely replaced gas proportional counting in most laboratories, is the liquid scintillation counting approach. With this method, the sample is first converted to benzene or a mixture of toluene and benzene. A scintillant such as PPO-POPOP is then added. When beta particles are released by the decaying radiocarbon, a pulse of light is emitted which is then detected and converted to an electrical signal using a photo multiplier tube. Liquid scintillation counting requires smaller samples and hence, it is preferable to gas proportional counting.

A reference standard which has a modern day radiocarbon activity is used by all laboratories in order to make the determinations comparable. The main standard has been the Oxalic Acid I sample maintained at the National Institute of Standards and Technology in Maryland, USA. The sample was derived from sugar beet in 1955. When stocks of Oxalic Acid I began to run low, a sample of Oxalic Acid II was obtained from beet molasses from France in 1977. The radiocarbon activity in Oxalic Acid II is very close to that of Oxalic Acid I. It is also pertinent to note that radiocarbon ages are presented in years before present (BP), present being 1950.

Errors encountered in beta counting include practical errors associated with the various laboratory procedures as well as statistical errors related to the decay process. Age calculations provide these uncertainties when reporting data. Dating using beta counting is sometimes referred to as conventional radiocarbon dating.

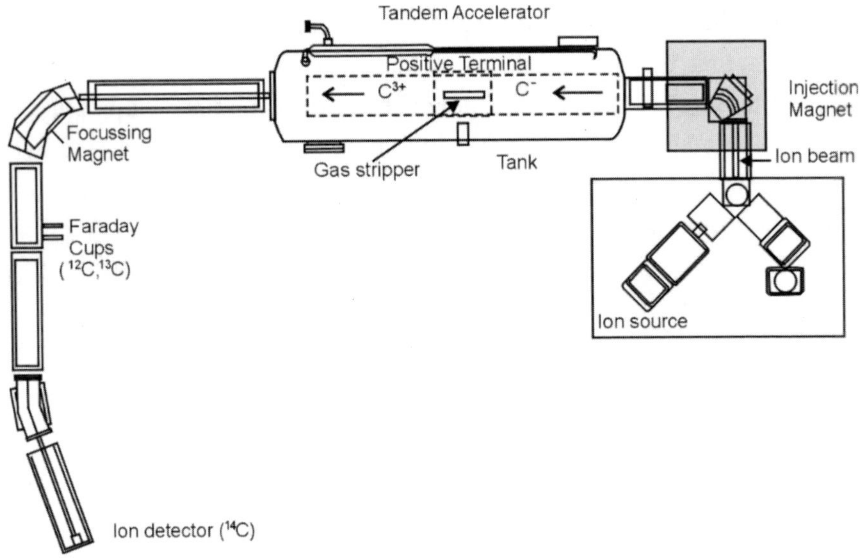


Figure 8. Basic configuration of a tandem accelerator mass spectrometer used for detecting and counting of radiocarbon atoms.

### *Accelerator Mass Spectrometry*

As outlined above, regular mass spectrometers cannot be used to detect $^{14}C$ isotopes in carbon because they cannot differentiate between the radiocarbon and other atoms of similar weight such as $^{14}N$. When isotopes of atoms are accelerated to high speeds using an accelerator, however, it becomes possible to differentiate between the small differences between, for instance, $^{14}C$ and $^{14}N$. AMS was developed in the 1980s by combining particle accelerators and mass spectrometers (Figure 8).

AMS measures the relative proportions of $^{14}C$ to $^{12}C$ and $^{13}C$. As a result, it is sometimes referred to as isotope ratio spectrometry (Walker, 2005). To determine the age, the ratio obtained from the sample is compared with that from a standard that has a known $^{14}C$ content. Typically, to reach the high speed required in AMS analysis, tandem acceleration is used. Before introduction to the spectrometer, samples are converted to graphite or some other source of $CO_2$ after which they are mounted onto a metal disc. By firing positively charged cesium ions ($Cs^+$) at the sample, negatively charged carbon ions ($C^-$) are produced. These are accelerated towards a positive terminal by passing through a stripper (Figure 8). The carbon ions lose three or four electrons to emerge from the stripper as positive ions ($C^{3+}$ or $C^{4+}$ depending on electrons lost). These are accelerated once again as a result of repulsion from the positive terminal and are then directed through a focusing magnet where deflection by mass occurs. Faraday cups can be used to measure $^{12}C$ and $^{13}C$ whereas ion detectors measure $^{14}C$ (Figure 8). To determine a radiocarbon age, a standard sample of known age such as oxalic acid from the US National Bureau of Standards is also run through the AMS and the ratios of $^{14}C$ to $^{13}C$ or $^{12}C$ that are obtained are compared to those of the sample of unknown age. As in beta counting, the uncertainties associated with the AMS age include errors arising from statistical uncertainties in the decay of $^{14}C$, as well as errors from laboratory measurements and from random and systematic sources (Walker, 2005).

A major advantage of AMS dating compared to beta counting is in the size of the samples that is required to make a successful measurement. Whereas beta counting typically requires 1-2 g of carbon, AMS dating can be done using about 1 mg of carbon. Hence, much smaller samples can be dated using AMS. AMS dating can also be completed in a few hours whereas beta counting requires several days or weeks of counting. Lastly, with AMS dating, multiple samples (up to 50 or more) can be analyzed in a single run, something that cannot be done with beta counting.

## Conventional Radiocarbon Age

A radiocarbon age determination is described as a conventional radiocarbon age (CRA) if it conforms to the following standards:

- it is calculated using a half –life of 5568 years (the Libby half – life)
- the modern radiocarbon standard used is either Oxalic acid I or II

- a correction is made for sample isotopic fractionation ($\delta^{13}C$) (explained below) to a normalized value of -25.0 ‰ with respect to the ratio of $^{12}C/^{13}C$ (the PDB standard).
- the calculation considers 1950 AD as the zero years datum.
- there is an assumption that all reservoirs of $^{14}C$ have remained constant indefinitely.

## Limits of the Radiocarbon Rating Timescale

For very old radiocarbon samples, dating becomes impractical because the residual activity of $^{14}C$ becomes so low that it is difficult to measure. Hence, ages beyond eight half-lives have largely been the limit when using regular radiocarbon dating approaches. However, efforts have been made to use isotopic enrichment which raises the ratio of $^{14}C$ to $^{12}C$ of a sample prior to measurement allowing ages of up to 75 ka to be reported (Stuiver et al. 1978). Other approaches have employed very sensitive counters with very large samples and reported ages of up to 60 ka (Behre and van der Plicht, 1992). In both instances, beta counting is used and very long counting times are required. For AMS measurements, specialized sample pretreatment protocols have been developed that have allowed ages of up to 60 ka to be obtained (Bird et al. 2003).

## Complicating Factors in Radiocarbon Dating: Sources of Error

When dating samples using radiocarbon methods, there are a number of assumptions that are made (Walker, 2005). The first assumption is that the ratio of $^{14}C/^{12}C$ in the atmosphere, biosphere and hydrosphere has remained the same throughout geological history. The second is that $^{14}C$ mixes rapidly throughout the atmosphere, biosphere and hydrosphere as soon as it formed such that the concentration is universally homogenous at any time. The third is that the ratio of $^{13}C/^{12}C$ isotopes in samples before the organism dies and after it has died remain unchanged and the only change in isotope ratio is as a result of the decay of $^{14}C$. The fourth is that the half-life of radiocarbon is known precisely. The final assumption is that the concentration of $^{14}C$ in nature is accurately known. Only the last two of these assumptions are thought to be correct. The first three are not and this results in a number of errors associated with radiocarbon ages (Walker, 2005) as is outlined below.

### *Contamination of Radiocarbon Samples*

Contamination errors occur when a sample is tainted by an older or younger sample such that the age obtained does not represent the true age of the sample. Such contamination could occur in the field, before the sample is collected. An example could be a sample in a lake where younger carbon is introduced by more recent biota. The $^{14}C$ activity obtained would be higher that what should be the correct level. Similarly contamination by older carbon, for instance from a disintegrating coal unit, could yield results suggesting a lower $^{14}C$ activity than should be the case, leading to erroneously older ages.

Apart from the possibility of natural contamination, sample contamination can also occur during the collection process, during transportation and in the laboratory. Hence, sample pretreatment procedures should try to eliminate all possibilities of contamination (Walker, 2005).

### *Isotopic Fractionation*

Biological pathways generally favor the lighter isotopes of carbon in any metabolic processes such that $^{12}C$ will be preferred to $^{13}C$ and $^{14}C$, and $^{13}C$ will be preferred to $^{14}C$. As result, the ratios of carbon isotopes in a growing plant will not be exactly identical to those of the ambient atmosphere since the $^{14}C$ will be slightly depleted in the plant. Also notably, different plant species have different levels of fractionation between the carbon isotopes. Hence, without correction, different plants that grew contemporaneously would exhibit different radiocarbon ages if they were to be dated simultaneously (e.g., Bowman, 1990).

In marine organisms the fractionation effect noted in plants is reversed because the $^{14}C$ is preferred such that the $^{14}C$ to $^{12}C$ ratio in sea water is higher than in the atmosphere. Marine organisms such as corals and molluscs that live in such water end up enriched in $^{14}C$ and would return radiocarbon ages that are slightly younger than their true age.

To correct for the isotopic fractionation effect, there is an assumption that enrichment of $^{14}C$ is twice that of $^{13}C$. This simplifies the procedure because $^{13}C$ and $^{12}C$ in a sample can be measured easily using a mass spectrometer. Once the ratio $^{13}C/^{12}C$ is determined, it is compared to that of a standard which is a Cretaceous belemnite carbonate from South Carolina (PDB limestone). The variation or deviation ($\delta^{13}C$) between the ratio of the standard and the sample will be given by the formula:

$$\delta^{13}C = \left[\frac{(^{13}C/^{12}C)}{(^{13}C/^{12}C)_{PDB}} - 1\right] x\ 1000 \qquad (8)$$

The value is expressed in parts per 1000 and the mean value for wood is -25 ‰. So once the $\delta^{13}C$ for a sample is calculated, it is normalized to that for wood before making a correction (Harkness 1979).

### *Marine Reservoir Effects*

Carbon dioxide is incorporated into sea water as dissolved carbonate. Thus, near surface waters in the oceans have $^{14}C$ levels that are close to those of the atmosphere. The $^{14}C$ in intermediate and deep ocean waters, however, is not easily replenished when it decays. Consequently, sea water will appear to be older than the present when dated using radiocarbon and such apparent ages in near surface waters range from 30 years in some seas to around 400-600 years in others. Deep oceans have longer residence times for ocean currents and apparent ages can reach thousands of years. Accordingly, radiocarbon ages obtained from marine organisms that extract carbonates to build their shells and other structures would need to be corrected for such reservoir effects before their ages can be analyzed.

A further complication with the reservoir effect is that in addition to the geographical variation, it also varies with time, not least because ocean circulation varies temporally (Reimer et al. 2002; Walker, 2005). Efforts continue to try and find solutions to the problem and marine reservoir databases have been compiled to help with the correction (e.g., Reimer and Reimer, 2001; Reimer et al. 2002; Butzin et al. 2016). It is important to point out that the reservoir effect has also been observed in freshwater (e.g., Philippsen, 2013).

### *Temporal Variations in the Atmospheric Production of $^{14}C$*

An aspect that is now well established is that the production of $^{14}C$ in the atmosphere has not been constant throughout geological history (e.g., de Vries, 1958; Stuiver and Suess, 1966). It is one of the main reasons radiocarbon years do not correspond to calendar years. These long-term variations, with periods of tens of thousands of years, are thought to be related to changes in the cosmic ray flux driven by temporal changes in the earth's magnetic field. Variations in the sun's activity have also been invoked as causes (e.g., Stuiver et al. 1991; Beck et al. 2001; Walker, 2005; Cauquoin et al. 2014; Reimer et al. 2013). The de Vries effect is a term that describes shorter term fluctuations in the atmosphere's $^{14}C$ content (Damon and Peristykh, 2000). These shorter oscillations have periods in the range of decades to hundreds of years.

Anthropogenic influence has also had an effect on the atmospheric concentration of $^{14}C$ over the last 250 years or so. This mainly arises from the combustion of fossil fuels which have released large quantities of $^{12}C$ and is referred to as the industrial effect. Since the period after the Second World War, the testing of thermonuclear devices had the opposite effect of adding $^{14}C$ into the atmosphere.

All these variations in the atmospheric concentration of $^{14}C$ over time necessitate that a correction is made before radiocarbon years can be compared with calendar years since the discrepancies can be significant. For instance radiocarbon ages obtained from fossil corals of around 15 $^{14}C$ ka BP have been shown to be around 3 ka (calendar years) younger than their true age and at 20 $^{14}C$ ka BP, the gap widens to about 4 ka (calendar years) (Stuiver et al. 1998; Walker, 2005). In essence, radiocarbon years are not equal to calendar years and when presenting them, the standard practice is to identify them with the units '$^{14}C$ years BP,' where BP denotes 'before present.' As outlined below, calibration of radiocarbon years is performed to convert the radiocarbon scale to the calendar scale so that such comparisons can be made.

## Radiocarbon Calibration

The lack of correlation between radiocarbon years and calendar years was first noted when radiocarbon ages were compared to ages determined using tree ring dating or dendrochronology. Since then, efforts have been invested in developing methods for converting $^{14}C$ years to calendar years. Dendrochronology has played a major role in this effort since it is possible to determine the ages of some trees (e.g., oak) using their rings as well as their $^{14}C$ age using the carbon content. As a result, dendrochronological calibration of radiocarbon has been conducted back to more than 10 ka PB. Other materials used for calibrating terrestrial radiocarbon ages include plant macrofossils from varved lake sediments and speleotherms (e.g., Reimer et al. 2013). The results are presented as calibration curves (e.g., IntCal04 (Reimer et al. 2004), IntCal09 (Reimer et al. 2009) and IntCal13 (Reimer et al. 2013)) which are used for correcting terrestrial radiocarbon ages. Curves are also prepared for marine radiocarbon ages (e.g., Marine09 (Reimer et al. 2009) and Marine13 (Reimer et al. 2013)). Calibrated radiocarbon years are denoted 'cal years BP' to distinguish them from uncalibrated $^{14}C$ years BP.

The Intcal 13 curve (Reimer et al. 2013) is the latest internationally ratified calibration curve and it can be used to calibrate radiocarbon ages in the range 0-50 cal ka PB. The first 13.9 cal ka BP are calibrated using tree rings (for the Northern Hemisphere) after which plant macrofossil data from Lake Suigetsu are used for the rest of the scale (Reimer et al. 2013). Gaps in the record are supplemented by comparing to data from other sources, including records from speleotherms as well as from marine sources.

## What Can Be Dated Using Radiocarbon?

Radiocarbon dating can theoretically be used to date any material that contains $^{14}C$ levels that are detectable and has functioned as a closed system such that no additional carbon isotopes have been gained or lost by any other means apart from the radioactive decay of $^{14}C$. In natural settings relevant to earth sciences, such materials are usually organic remains from organisms that incorporated $^{14}C$ during their lifetime. Materials that are commonly used include plant macrofossils, peat, organic lake mud, marine microfauna and microflora, shell and bone. Table 4 outlines some characteristics of these materials and more details can be found in Walker (2005).

**Table 4. Terrestrial materials that can be dated using radiocarbon**

| Material | Dated Component | Comments |
|---|---|---|
| plant macrofossils | wood, seeds, leaves, peat | can be well preserved with minimal contamination |
| charcoal | charcoal | can be well preserved |
| organic lake mud | organic material in sediment | susceptible to contamination by older carbon |
| mollusc shell | calcium carbonate | susceptible to contamination |
| bone | hydroxyapatite | contamination likelihood - continuous carbon exchange |
| bone | collagen | less susceptible to contamination but degrades rapidly |
| soil | organic fraction | susceptible to contamination |

## Constraining the Last Deglaciation Using Radiocarbon Dating

Radiocarbon dating has been used to constrain deglaciation by providing ages that approximate a time period by which the landscape had become ice-free (e.g., Mickelson and Colgan, 2003). The rationale for employing radiocarbon dating for constraining deglaciation is that, the appearance of organic material on the postglacial landscape is an indication that ice sheets had now receded and organic material, whether floral or faunal, had become established. The main problem of this approach, depending on the type of organic material used, is that, following the retreat of an ice sheet, the climate has to ameliorate enough to allow biotic communities to flourish. Hence, the appearance of the plants or animal life on the postglacial landscape does not necessarily coincide with the retreat of the ice sheet. Compounding the problem is that it is difficult to estimate how long it takes for organic material to colonize the landscape after the ice sheets have receded (e.g., Clayton and Moran, 1982; Fisher et al. 2009). Using modern analogues, Fisher et al. (2009) have argued that the period between deglaciation and the appearance of vegetation can be as short as three decades. That, however, is probably on the lower end of estimates (Munyikwa et al. 2017). This means that radiocarbon ages from postglacial terrestrial organic material simply provide minimum age estimates by which the ice sheet would have retreated.

Despite the concerns outlined above, in the absence of an alternative chronometer, radiocarbon dating can provide invaluable minimum age constraints for the emergence of an ice-free surface on a previously glaciated landscape. In North America, Dyke et al. (2001) compiled radiocarbon ages from various studies and used them to constrain the growth and recession of the Laurentide and Innuitian Ice sheets following the LGM. Radiocarbon ages obtained from sources such as basal lake sediment with low organic content and samples that were likely contaminated by older carbon (e.g., through the hard water effect or otherwise) were excluded from analysis. Accordingly, ages used in the reconstruction comprised chronologies from peat, gyttja, wood and other plant materials from lakes and depressions. Ages from bone of mammals such as horses and mammoth were also included. For the Innuitian Ice Sheet and the eastern margin of the LIS, marine carbonate shells were employed following correction for the reservoir effect (Dyke et al. 2001). Using the data, Dyke et al. (2002, 2003) created deglaciation models that show the retreat of North American ice sheets between the termination of the LGM and the start of the Holocene.

In western Canada, Fisher et al. (2009) used more than 70 radiocarbon ages obtained from postglacial lakes in northeast Alberta in an effort to constrain the timing of the retreat of the LIS from the region. Sampling from kettle lakes was avoided (Fisher et al. 2009) since these are formed by processes that persist after the main ice sheets have melted and, hence, would make it difficult to ascertain the timing of deglaciation. Small lakes that formed in bedrock were targeted instead. Samples were collected by coring close to the lake margin in an effort to maximize collection of terrestrial macrofossils. Whenever practical, lakes close to moraine ridges were cored in an effort to obtain temporal constraints for the ice sheet margin's retreat. Sampling was particularly directed at wood or fragments of wood in order to minimize the possibility of contamination. However, other materials that were analyzed included charcoal, seed pods, peat, moss and other types of plant fragments (Fisher et al. 2009).

Multiple other studies have employed radiocarbon ages to constrain the retreat of the LIS to the extent that Mickelson (2001) has previously described it as the best dated ice sheet margin in the world. The studies are far too numerous to mention here and can be found in reviews that include Dyke et al. (2001), Arnold (2002), Mickelson and Coglan (2003), Dyke (2004, 2005), and Hill (2006).

In Europe, Lasberg and Kalm (2013) compiled ages obtained using radiocarbon as well as luminescence and cosmogenic exposure dating methods and used the chronologies to place constraints on the growth and deglaciation of the Scandinavian Ice sheet during the Late Pleistocene (Late Weichselian). The radiocarbon samples were obtained from materials that included peat, moss plant detritus and mammoth teeth (Lasberg and Kalm, 2013). Within the same region, Raukas et al. (2010), have also employed radiocarbon data in conjunction with OSL chronologies to constrain the Late Pleistocene deglaciations of the south-eastern Baltic region.

Marine fauna contained in deglacial muds in northern Britain were dated using radiocarbon methods and the chronologies obtained used to constrain the Late Pleistocene deglaciation of the British Isles (McCabe, 1996). Marine fauna in marine muds that overlie or are interbedded with deglaciation features were similarly dated using radiocarbon methods and used to constrain the retreat of the British-Irish Ice Sheet (McCabe et al. 2005).

As with the LIS in North America, studies that have employed radiocarbon ages to constrain the chronology of European ice sheets are far too many to list in this chapter. However, Hughes et al. (2016) have presented an updated chronological database and time-slice reconstruction of the last Eurasian ice sheets which demonstrates the invaluable role played by both radiocarbon as well as OSL and cosmogenic nuclide dating.

## Limitations of Using Radiocarbon Ages to Constrain Deglaciation

Apart from issues related to the need for calibration of the $^{14}C$ scale and the susceptibility to contamination that may affect radiocarbon chronologies as outlined above, the difficulty of determining the time lag between the retreat of an ice sheet and the establishment of biotic communities following deglaciation is also an important factor when working with radiocarbon ages. Radiocarbon dating that has been performed in western Canada offers pertinent illustration (e.g., Dyke et al. 2001; Fisher et al. 2009). When collecting samples in the region, wood is particularly preferred. However, it is thought that following the retreat of the LIS from western Canada, the first vegetation to develop in the region was tundra (e.g., Dyke, 2005), which comprised mostly grasses. Wood only appeared later when boreal vegetation colonized the region. Nonetheless, it has not been possible to determine the time lag between ice sheet retreat and the appearance of woody vegetation. While some have argued that this gap was brief (e.g., Fisher et al. 2009), it has been noted by others (e.g., Munyikwa et al. 2017) that, following deglaciation, the soils in western Canada would have been rich in lime such that a period of decalcification through leaching would have been required before woody vegetation could thrive. As a result, the time lag between deglaciation and when the wood became established may have been protracted. All in all, this uncertainty suggests that constraints obtained from radiocarbon ages should be treated as conservative minimum estimates for deglaciation that, in some cases, could underestimate the true timing of ice sheet retreat significantly. However, in settings where the dated material is known to have been in place soon after the retreat of the ice sheet (e.g., McCabe et al. 2005), the issue of the time lag becomes irrelevant.

## VARVE CHRONOLOGY

### Introduction

Varve chronology is a dating method that uses the addition of annual layers to a sediment sequence as a measure of time. Such sediment layers would typically be added in a lake or in the sea. In this chapter, however, our focus will be on varved sediments in lakes that occur on continental landscapes and can be used to constrain the timing of deglaciation. By comparing varve thicknesses from one site with that of varves from another site, it is possible to correlate contemporaneous varve units. In this manner, varves from different sites can progressively be linked to produce a continuous varve sequence. As will be noted below, the varve sequences form floating chronologies that need to pegged to a certain period using another absolute age chronometer to serve as an anchor point so that the annual layers can be used as measures of calendar time (Walker, 2005; Zoilitschka et al. 2015).

As a tool for dating deglaciation, varve chronology has a special place because it was the first absolute dating method to be used for determining the timing of ice sheet recession, well before the introduction of radiogenic dating methods. The pioneering studies were conducted by the Swedish geologist Gerard De Geer who was trying to use varves to ascertain the chronology of the retreat of the Fennoscandian Ice Sheet after the LGM (Walker, 2012; Francus et al. 2013). Indeed, at the Geological Congress of 1910 that was held in Stockholm, De Geer (1912) proposed that the term geochronology only be used when referring to varves, since he considered all other dating methods of the time to be too imprecise. De Geer noted the annual character of lake sediments in glacial lakes that formed in front of the retreating Fennoscandian ice sheet as it retreated northwards following the termination of the LGM. He therefore theorized that by tracing the layers in the lakes northwards, it would be possible to link the units from different lakes to generate a continuous chronology of the otherwise floating timescale.

## Basic Principles of Varve Chronology

Varved sediments in lacustrine environments occur in various forms. Clastic varves comprise sequences of interbedded layers of coarser grained siliciclastic deposits with intervening finer grained sediments such as silts and clay. Such sediments are sourced from outside the lake and are transported by fluvial processes or subaerially by wind. Proglacial and periglacial lakes are particularly favorable sites for this type of sediment as the pervasive physical weathering associated with gelifraction and the absence of vegetation cover provides a plentiful supply of easily erodible clastic sediments (Zolitschka, 2015). Conditions during spring and summer promote different depositional conditions compared to fall and winter (Walker, 2005; Zolitschka et al. 2015). The ice-free conditions that characterize spring and summer seasons allow coarser grained sediments to be deposited in the lake. During the winter, however, when the lakes typically freeze over, fine gained silts and clays settling from suspension are emplaced above the coarser material. Within a given year, therefore, the sedimentary sequence would comprise a couplet made up of a coarse unit and a fine grained unit (Walker, 2005).

In some settings, clastic-organic (or clastic-biogenic) varves, also referred to as mixed varves, are formed when alternating layers of clastic sediments are overlain by organic materials. These develop when floods wash clastic sediments from the lake's watershed into the lake during spring. In the summer, organic material such as plant and algal detritus are then deposited above the clastic sediments. During the winter when the lake is possibly frozen and input into the lake is minimal, fine grained organic matter may be deposited to cap the annual sequence (e.g., Tiljander et al. 2002, 2003; Ojala and Alenius, 2005; Haltia-Hovi et al. 2007).

Organic (or biogenic) varves consist solely of organic material or biomineralisation and are the product of accentuated seasonal variations in climate that lead to significant changes in the availability of nutrients for organic productivity (Zahrer et al. 2013). In such lakes, light-colored organic material accumulates during the summer as a result of elevated productivity. In the winter, however, cooler temperatures are accompanied by slower accumulation rates that result in a thinner and darker layer (e.g., Walker, 2005; Saarni et al. 2015).

Variations in calcium carbonate (or endogenic) precipitation between seasons can lead to color differences that are annually repetitive between layers to give calcareous varves (Walker, 2005; Zolitscha, 2015). Similarly, a rise in diatom blooms during the summer could result in diatom-rich varves while layers that arise from variations of iron oxide precipitation between seasons produce iron-rich varves (Walker, 2005).

Annual laminations could also develop in ocean sediments to give varves (e.g., Matthews, 2012; Johnson et al. 2015). Where glaciers terminate in the sea, for example, seasonal variations in sediment released by the glacier could yield annual pairs of diamict and laminated muds that form glaciomarine varves. Diamict would be deposited in the winter months when glacier melting is reduced. During the summer, however, the melting glacier would release suspended sediment to form the laminated muds (Walker, 2005).

Variations that occur in ocean productivity such as those related to annual overturning that affects ocean nutrients could result in laminated sediments. Layers that contain high levels of microflora such as diatoms would be lighter in color compared to diatom-poor layers. Such couplets are referred to as marine varves.

## Counting and Analyzing Varves

Varve chronology entails identifying the annual varve structure after which the individual couplets may be enumerated. While the counting can be performed in the field in some cases, the most common practice today is to conduct the counting in a laboratory setting. Sampling is carried out by coring (e.g., Zahrer et al. 2013; Saarni et al. 2015), or by collecting monoliths from open profiles. In the laboratory, the annual varve pairs can be identified with the unaided eye, or alternatively by using light microscopy. More recent studies, however, commonly use thin section methods (e.g., Zahrer et al. 2013; Saarni et al. 2015), scanning electrom microscopy (SEM), photography and digital image analysis approaches as well as X-ray densitrometry to help identify the varves (e.g., Ojala and Alenius, 2005; Zolitschka et al. 2015). Such techniques enable very fine varves to be analyzed. With SEM analysis, adding an energy dispersive spectroscopic analyser enables the elemental composition of laminae to be examined (Dean et al. 1999). Micro-Xray fluorescence (μ-XRF) can also be used to provide quantitative elemental analysis for individual laminae (e.g., Martin-Puertas et al. 2017)

## Applications of Data from Varves

Varves are important repositories of paleoenvironmental data (Walker, 2005). Their analysis allows a precise chronological framework to be established alongside markers of environmental change. Accordingly, varve chronology has been applied in multiple studies that have examined climate change (e.g., Lucke et al. 2003; Ojala and Alenius, 2005; Martin-Puertas et al. 2017), vegetation change (e.g., Brauer et al. 1999; Zahrer et al. 2013; Martin-Puertas et al. 2017; Slowisnki et al. 2017), land-use histories (Zolitschka et al. 2003) as well as geomorphic change (e.g., Balco and Schaefer, 2006; Smith et al. 2013; Schlolaut et al. 2014). In geomorphic studies, the chronology from varves has enabled quantitative analysis to be performed to elucidate rates of erosion from catchments as well as rates of sediment input in lakes (e.g., Schlolaut et al. 2014; Laroque-Tobler et al. 2015). In more modern settings, pollution by heavy metals, eutrophication and the effects of acid rain have been studied using chronology from varves (Walker 2005). Additionally, as outlined above, varve chronology has been applied in the calibration of longer term radiocarbon ages, beyond the range of dendrochronological records (e.g., Nakagawa et al. 2012; Reimer et al. 2013). In other applications, data from varves have been used to reconstruct variations in Earth's magnetic field from the past which can also be employed in core-to-core correlation between depositional sites in varve sequences (Zolitschka et al. 2015). In this chapter however, our focus is on the use of varves as constraints for deglaciation.

## Constraining the Last Deglaciation Using Varves

The contributions of varve chronology to constraining Northern Hemisphere deglaciation after the LGM are significant. As pointed out earlier, deposition of varves in proglacial and glacial lakes provide both a physical location of the ice sheet margin as well as a chronological tool for approximating the timing. Zolitska et al. (2015) provide the global distribution of published lacustrine varved sediment records from the Pleistocene and Holocene. The majority of these sites are in areas formerly covered by the LIS in North America and the Fennoscandian ice sheet in Western Europe. Many of the varve sequences have been used to constrain deglaciation from the respective sites. In Europe the best known of these are the Swedish Varve Chronology series. In the USA, well studied regions include the varves of New England that formed the southeast margin of the

LIS as well as those in the Lake Superior region. In Canada, the Lake Ojibway varves are worth mentioning.

In Sweden, De Geer's early efforts aimed to create a varve chronology for the retreat of the Fennoscandian Ice Sheet as it receded to the north and to the west. The east and southeastern margins of the ice sheet were bounded by the Baltic Ice Lake which was situated in what is now the Baltic Sea. In places where rivers debouched into the lake, clastic varves formed progressively adjacent to the ice sheet margin. Since De Geer's pioneering studies, the Swedish varve sequence has been revised and by linking records from individual sites, a master chronology of the last ca. 13-14 ka has been created though minor gaps are thought to exist in the sequence (Walker, 2005). Such breaks can be resolved by incorporating data from other sources such as pollen records or radiocarbon ages from varved clay sequences as well as Greenland ice cores (Walker, 2005).

In North America, the systematic study of varves was introduced by Ernst Antevs (1922, 1925, 1928, 1931) who had trained in Sweden under De Geer. Antevs' work in the Connecticut Valley led to the development of the New England Varve Chronology (NECV) series. Many decades later, however, the NECV was reformulated and renamed as the North American Varve Chronology (NAVC). The NAVC is a continuous sequence covering 5659 years extending from ca. 18.2 ka to 12.5 ka years BP. Hence, the timeline provides constraints for the last deglaciation of the region. The thickness of varves from New England was compared to Greenland ice core records and results suggest that, after ca. 15 ka, the climate in the North Atlantic was linked to processes in the southeastern region of the LIS. Ice sheet retreat also seemed to accelerate after 15 ka (Ridge et al. 2012).

In the Great Lakes region of North America it is thought that, as the LIS receded northwards from the area now occupied by Lake Superior, it opened up an eastern outlet for glacial Lake Agassiz which lay to the west. This allowed glacial meltwater as well as discharge from Lake Agassiz to spill into Lake Superior, depositing sequences of rhythmites in the process. Using time series analysis of the thickness measurements and carbonate data on sediment cores, the rhythmites were shown to be varves. By creating a chronology using the varves it has been shown that the recession of the LIS from the Lake Superior region occurred between ca. 10.7 and 8.9 cal ka BP (Brekenridge, 2007). Varve formation ultimately terminated when the inflow was captured by glacial Lake Ojibway, which formed on the southern margin of the LIS as the last major proglacial lake of the last glaciation. Brekenridge et al. (2012) analyzed varves from glacial Lake Ojibway, an area previously studied by Antevs (1925) and concluded that the records there represent a total of 2.1 ka. There was no material in the varves that could be dated using absolute age methods to assign calendar ages. However, if the deposition of varves in Lake Ojibway is assumed to have begun immediately after the redirection of flow from Lake Superior into Ojibway, then varves in the lake could be constrained to ca. 10.6 – 8.5 cal ka BP (Brekenridge 2012). This would suggest that glacial Lake Ojibway ultimately drained around 8.5 ka when the ice dam from the waning LIS finally failed. By linking the varve sequences in the direction of ice sheet retreat, Brekenridge et al. (2012) were able to approximate retreat rates of the LIS as well as constrain epidoses of readvance.

There are numerous other studies incorporating varves that have been conducted in North America to constrain the retreat of the LIS (e.g., Rittenour et al. 2000; Ridge, 2003, 2004; Balco and Schaefer, 2006; Stone and Stone, 2012). In New England, for instance, while other chronometers such as cosmogenic nuclide and radiocarbon dating have also been used to constrain the retreat of the ice sheet margin, the chronology from varves is thought to be the most accurate (Balco and Schaefer, 2006). Hence, in most instances, these studies demonstrate the potential for data from varves to impart a high resolution and quantitative perspective to the reconstruction of deglaciations.

## Limitations of Varve Chronology and Future Prospects

A basic assumption in varve counting is that a single couplet of layers or laminations is formed in any given year. However, there are situations in which this assumption may be violated such that additional layers are formed. Instances where this may occur include lakes that mix twice a year (dimictic) where silt layers may be deposited during spring and autumn and organic layers during the summer and winter. Seasonally dependent precipitation of iron may also introduce additional color banding (Walker, 2005). Multiple laminations could also occur in environments where there are diurnal changes in sediment deposition such that coarse sediments are emplaced during the day and fine grains overnight. This may be the result of episodic currents that are generated by local winds or other stochastic weather events such as floods or multiple periods of snow melting in a given year (Walker et al. 2005).

The elimination of varves as a result of bioturbation, erosion, or other geomorphic processes could also lead to complications in varve counting (Walker, 2005; Zolitschka et al. 2015). Varves are best developed in deep lakes with flat central basins as the steep margins are susceptible to sediment slumping and mass wasting processes (Sauerbrey et al. 2013). Deeper lakes are also less susceptible to bioturbation as bottom waters tend to be more anoxic (Zolitschka et al. 2015). Excessively high or very low sedimentation rates could make it difficult to identify varve couplets, which would introduce counting errors (Zolitschka, 2015). In addition to errors that can occur during the development of varves, systematic errors could also result from incomplete core recovery during sampling, causing gaps in the sediment record.

In all instances, such cases of additional or missing layers could be addressed by collecting multiple cores from a lake and comparing the records. Similarly the use of markers that are well formed as calibration points between cores could be a useful safeguard. When trying to compare cores from different sites for purposes of correlation, however, additional laminations or missing laminations could pose significant impediments.

Despite the above limitations, varve chronology is a high-resolution dating method that has been indispensable to the study of the late Quaternary. The dating method provides an accurate temporal scale to a range of environmental and geomorphic variables. New advances in the study of varves has now made it possible to monitor and model modern varve systems, leading to an improved insight into their formation. Improved instrumentation has the capacity to augment this understanding further. Ultimately the data acquired from such studies should lead to better climate models (e.g., Braconnot et al. 2012). Suggestions have been put forward for a unified set of protocols for metholodigal approaches in varve studies, as well as for reporting of dating results, including uncertainties (e.g., Brauer et al. 2014; Zolitschka et al. 2015). Such initiatives should enhance the quality of the data and facilitate the harmonization and integration of chronologies from different dating methods that are used in global studies of climate change (Zolitschka et al. 2015).

## Other Dating Methods that Could Be Used to Constrain Continental Deglaciation

In theory any other Quaternary dating methods that could be used to approximate time that has elapsed since an ice sheet or glacier receded from a given landscape could be employed to provide constraints for deglaciation. For the Late Pleistocene, these could include dating methods such as tephrochronology (e.g., Beget, 1984; Thornalley et al. 2011) which employs tephra layers constrained using another absolute age method. Relative dating of surfaces using soil chronosequences, which are suites of soils that develop under similar conditions, has been proposed for use on post-incision surfaces such as postglacial landscapes (e.g., Harden, 1982; Hugget, 1998). At best, however, pedogenesis can only be used as a measure of time locally since the development of soils is influenced by local factors that include biotic and climatic aspects (e.g., Vidic and Lobnic, 1997; Schaetzl et al. 2006). Nonetheless, the primary focus of this chapter is on absolute age methods that are routinely used for constraining deglaciation. Hence, relative dating methods will not be discussed further.

# Constraining the Late Pleistocene Retreat of the Laurentide Ice Sheet from Western Canada

## Introduction

To illustrate the multiple considerations that could go into the selection of a dating method to assign a chronological framework to the retreat of an ice sheet, this section gives a brief outline of efforts made to constrain the Late Pleistocene retreat of the LIS from western Canada. At the height of the LGM, between ca. 24 and 20 ka ago, the LIS expanded to cover most of Canada (e.g., Dyke et al. 2001). On its western margin, the LIS merged with the Cordilleran Ice Sheet (Figure 9a). Using radiocarbon chronologies from areas formerly covered by the LIS, Dyke et al. (2001, 2002, 2003) have created glacial retreat models for North America. The radiocarbon data, however, are not evenly distributed throughout the region. On the eastern and southern fronts of the LIS, the radiocarbon data are relatively sound and the progressive ice sheet margins are thought to be well-constrained (e.g., Dyke et al. 2002). For western Canada, however, limiting data for ice sheet margin positions for the period 20 -11 ka are scarce. Nonetheless, the chronology suggests that the LIS began to retreat from the province of Alberta around 20 ka and by ca. 11 ka, most of the region was ice-free (Figure 9). Beyond this broad framework, it has not been possible to provide a detailed reconstruction of progressive positions of the LIS between 20 and 11 ka. Thus, current radiocarbon derived reconstructions should only be seen as tentative (Munyikwa et al. 2011a, 2017). Furthermore, large parts of formerly glaciated regions of western Canada do not have any moraines that could be dated using cosmogenic nuclides. Hence, alternative chronometers are needed.

Luminescence dating of postglacial eolian sediments that occur in Alberta has been identified (Wolfe et al. 2004) as a potential dating method that could be used for constraining the retreat of the LIS from the region. This is because there are more than 120 postglacial eolian dune fields in the area, the majority of which are found in Alberta (Wolfe et al. 2007;

Halfen et al. 2015) (Figure 9c). Source sediments for the eolian deposits are thought to be glaciolacustrine and deltaic sediments associated with the retreating ice sheet margin. Because the regional topography slopes to the northeast, as the LIS retreated northeastwards, large proglacial lakes were progressively dammed by the ice sheet margin. With continued ice sheet recession, the glacial lakes drained leaving behind fine grained sediments that were subsequently mobilized by wind to form the dune fields (e.g., Wolfe et al. 2004, 2007; Munyikwa et al. 2011a; 2017). Dune formation is thought to have been driven by three main factors: low soil moisture, abundant sediment supply and strong winds on the periphery of the ice sheets. Subsequently, as the climate continued to improve behind the retreating LIS, vegetation colonized the postglacial landscape, stabilizing the dune fields. In central and northern Alberta, where boreal vegetation developed following deglaciation, the dune fields have been preserved as fossil postglacial landforms. Hence, by dating the eolian dunes using luminescence methods, minimum age constraints can be assigned to the deglaciation of the region (Wolfe et al. 2004; Munyikwa et al. 2011a, 2017).

Assigning a deglaciation chronology to western Canada is important for a number of reasons. First, the region is at the center of the proposed Late Pleistocene ice-free corridor (Johnston, 1933) that may have been used by the First Americans to spread to the mid- and low-latitude regions of the continent by ca. 15 ka. An accurate deglaciation chronology would allow evaluating whether that was possible (Munyikwa et al. 2011a; 2017). Second, the LIS retreat allowed genetic exchanges to take place between floral and faunal communities that may have been separated by the ice sheet between Beringia and regions south of the ice sheet. Hence, an improved timing of deglaciation would allow an accurate determination of when the exchanges would have been possible. Third, it has previously been proposed that meltwater released by the LIS triggered the Younger Dryas stadial event by modifying thermohaline circulation in the North Atlantic (e.g., Broecker et al. 1989).

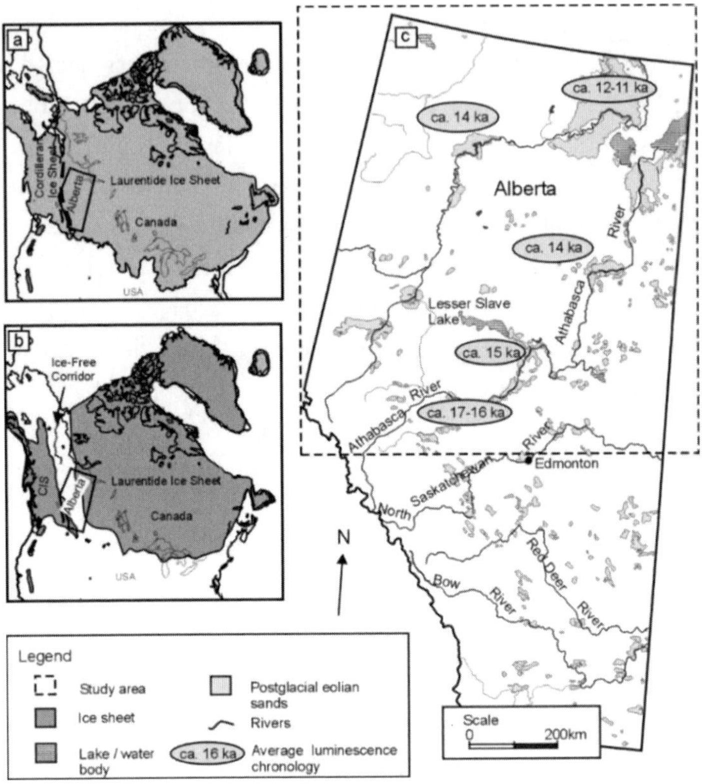


Figure 9. (a) Extent of the Laurentide Ice Sheet in North America at the LGM (after Dyke et al. 2001). The province of Alberta was entirely covered by the LIS (with the exception of a few isolated elevated points) (b) The conceptual ice-free corridor lying between the Cordilleran Ice Sheet (CIS) and the LIS that was proposed by Johnstone (1933) to have been used as an inland migratory corridor by the First Americans. Note that the corridor passes through Alberta. (c) The province of Alberta after the retreat of the LIS showing the distribution of postglacial dune deposits. The study area is in central and northern Alberta. Average luminescence ages for eolian dune initiation are indicated.

An improved deglaciation chronology would facilitate a determination of whether the meltwater was routed to the ocean eastwards to North Atlantic or northwestwards to the Arctic Ocean (Munyikwa et al. 2011a). Finally, deglaciation chronologies would be invaluable for calibrating geophysical models that look at regional mantle rheology and postglacial isostatic rebound as well as postglacial sea level change (e.g., Tarasov and Peltier, 2006).

## Methods

Luminescence dating samples were collected from eolian dunes in central and northern Alberta by excavating pits to expose fresh sediment (e.g., Wolfe et al. 2004; Munyikwa et al. 2011a, 2017). Where open sections were available, sampling was performed by extracting sand from freshly cleaned profile faces. In other places, samples were extracted by coring vertically with a powered auger. In all instances, samples were retrieved in opaque containers and immediately sealed to avoid exposure to light and loss of moisture. At the lab, the samples were processed using standard luminescence dating procedures to isolate sand sized grains of either K-feldspar (Wolfe et al. 2004; 2007) or quartz (Munyikwa et al. 2011a, 2017). The separates were then analyzed using IRSL (Wolfe et al. 2004, 2007) or blue OSL (Munyikwa et al. 2011a; 2017). Dose rates were ascertained using elemental concentrations of U, Th and K in the samples after which dose contributions were calculated using conversion factors of Guerin et al. (2011). The total dose rate included cosmic ray contribution which was calculated using the method of Prescott and Hutton (1994). Dose rate attenuation resulting from interstitial moisture was also taken into account (e.g., Aitken, 1998).

## Tentative Findings

The chronology of eolian deposition in central and northern Alberta suggests that the dunes in the region were initiated time-trangressively, starting at around 17-16 ka in the southwestern part of the study area and progressed northeastwards, ending around 12-11 ka in the northeastern part of the province (Figure 9c). This trend is consistent with the retreat pattern of the LIS. The data also indicate that by 15 ka, central Alberta was free of ice, suggesting that the ice-free corridor, proposed by Johnstone (1933) as having been used by Paleoindians to spread to areas south of the ice sheets, may have been open earlier than some studies (e.g., Mandryk et al. 2001; Madsen, 2015) have suggested. More dating will need to be conducted north of the area already studied to determine if the entire corridor was open early enough for the first Americans to have used it as an inland migratory corridor. Similarly, a chronology for the opening of the northern end of the corridor would enable investigators to ascertain when genetic exchanges may have become possible prior to ca. 14 ka between areas previously separated by the LIS. With regards to meltwater routing, dating more samples from northeast Alberta should help establish whether the drainage of meltwater to the Arctic Ocean around ca. 12.7 ka was possible such that it triggered the Younger Dryas stadial event. Overall, however, the retreat chronology of the LIS reconstructed to date from luminescence dating does not preempt any of the possibilities outlined above (Munyikwa et al. 2011a).

## CONCLUSION

Geochronologic methods that can be used to develop absolute age frameworks for Late Pleistocene ice sheet retreat should fulfill two main criteria. First, the methods should be capable of dating geomorphic features that are produced by the glacier or, alternatively, date landforms or materials that appear on postglacial landscapes soon after the deglaciation. Second, the methods should have a dating range that spans the last ca. 5 – 30 ka which is the later part of the Late Pleistocene. Accordingly, there are generally four main dating methods that can be used for developing temporal frameworks for the last deglaciation: cosmogenic nuclides dating, luminescence dating, radiocarbon dating and varve chronology. In theory, other Quaternary dating methods could be employed to constrain deglaciation. However, the four methods outlined above are the most routinely used techniques.

Of the four geochronologic techniques, the choice of which method to use in any given study area will depend on the deglaciation features that are available. An ideal setting would be where a range of features are present, making it possible to employ multiple dating methods and come up with a robust chronology, should the results be congruent (e.g., Balco and Schaeffer, 2006). More often than not, however, it is the case that only one method can be employed satisfactorily.

With regards to future prospects for dating deglaciations, all dating methods discussed above are undergoing improvements as a result of both technical advances in the methods themselves as well as from improved understanding of the processes involved in deglaciation. As a result, a number of improvements should stem from these advances. First, chronologies that are more accurate should lead to a better understanding of the driving forces behind deglaciation (e.g., Ridge et al. 2012). Second, better deglaciation chronologies enable researchers to correlate environmental evolution, including vegetation, with climate changes more accurately (e.g., Williams et al. 2004; Dyke, 2005). Third, an enhanced deglaciation chronology should help produce geophysical models of regional postglacial isostatic adjustment (e.g., Tarasov and Peltier, 2006) that are better calibrated. Fourth, extensions in upper age limits of some of the dating methods could make it possible to constrain deglaciation events beyond the last deglaciation with improved resolution. Progress in all these areas would potentially have a range of benefits not just in geology and earth sciences but also in biology as well as in archaeological sciences. This would particularly be the case in North America where the first Americans are thought to have arrived after the LGM, with their dispersal possibly being facilitated geographically by the recession of the ice sheets (Munyikwa et al. 2011a; 2017).

## ACKNOWLEDGMENTS

Gratitude it extended to Susan Fryters for providing editorial support during the preparation of this chapter.

## REFERENCES


Aitken, M. J. (1985). Thermoluminescence Dating. London: Academic Press.

Aitken, M. J. (1998). An Introduction to Optical Dating. Oxford: Oxford University Press.

Akçar, N., Yvuz, V., Ivy-Ochs, S., Kubik, P. W., Vardar, M. & Schlüchter, C. (2008). A case for a downwasting mountain glacier during termination I, Verçenik valley, northeastern Turkey. *J. Quat. Sci.*, *23*, 273-285.

Alexanderson, H. & Murray, A. S. (2007). Was southern Sweden ice free at 19-25 ka, or were the post LGM glacifluvial sediments incompletely bleached? *Quat. Geochronol.*, *2*, 229-236.

Antevs, E. (1922). The recession of the last ice sheet in New England. Am. Geograph. Soc. Res. Series n., *11*, 120 p.

Antevs, E. (1925). Retreat of the last ice sheet in eastern Canada: Geological Survey of Canada, Memoir, *146*, 142 p.

Antevs, E. (1928). The last glaciation, with special reference to the ice sheet in northeastern North America. *Am. Geograph. Soc. Res. Series*, n. *17*, 292 p.

Antevs, E. (1931). Late-glacial correlations and ice recession in Manitoba. *Geological Survey of Canada Memoir*, *168*, 76 p.

Arnold, J. R. & Libby, W. F. (1949). Age determinations by radiocarbon content: checks with samples of known age. *Science*, *10*, 678-680.

Arnold, T. G. (2002). Radiocarbon dates from the ice-free corridor. *Radiocarbon*, *44*, 437-454.

Azizov, A. (1977). The influence of soil moisture on the resistance of soil to wind erosion. *Sov. Soil Sci.*, *9*, 105-108.

Bailey, R. M., Smith, B. W. & Rhodes, E. J. (1997). Partial bleaching and the decay form characteristics of quartz OSL. *Radiat. Meas.*, *27*, 124-136.

Balco, G. (2011). Contributions and unrealized potential contributions of cosmogenic-nuclide exposure dating to glacier chronology, 1990-2000. *Quat. Sci. Rev.*, *30*, 3-17.

Balco, G. & Schaefer, J. M. (2006). Cosmogenic-nuclide and varve chronologies for the deglaciation of southern New England. *Quat. Geochronol.*, *1*, 15-28.

Barker, S., Diz, P., Vautravers, M. J., Pike, J., Knorr, G., Hall, I. R. & Broecker, W. S. (2009). Interhemispheric Altantic seesaw response during the last deglaciation. *Nature*, *457*, 1097-1102.

Bartoli, M., Sarnthein, M., Weinelt, M., Erlenkeuser, H., Garbe-Schonberg, D. & Lea, D. W. (2005). Final closure of Panama and the onset of northern hemisphere glaciation. *Earth Planet. Sci. Lett.*, *237*, 33-44.

Beck, W. J., Richards, D. A., Edwards, R. L., Silverman, B. W., Smart, P. L., Donahue, D. L., Hererra-Osterheld, S., Burr, G. S., Calsoyas, L., Jull, A. J. T. & Biddulph, D. (2001). Extremely large variations of atmospheric $^{14}C$ concentration during the last Glacial Period. *Science*, *292*, 2453-2458.

Beget, J. E. (1984). Holocene glacier fluctuations near Glacier Peak, North Cascade Range, Washington. *Quat. Res.*, *21*, 304-316.

Behre, K. E. & van der Plicht, J. (1992). Towards an absolute chronology for the last glacial period in Europe: radiocarbon dates from Oerel, northern Germany. *Veg. Hist. Archaeobot.*, *1*, 111-117.

Belly, P. Y. (1964). Sand movement by wind. US Army Corps of Engineers. *Coastal Engineering Research Centre*, *Tech. Memo* No.1

Berner, R. A., Lasaga, A. C. & Garrels, R. M. (1983). The carbonate-silicate geochemical cycle and its effect on atmospheric carbon dioxide over the past 100 million years. *Am. J. Sci.*, *283*, 641-683.

Bickel, L., Lüthgens, C., Lomax, J. & Fiebig, M. (2015). Luminescence dating of glaciofluvial deposits linked to the penultimate glaciation in the Eastern Alps. *Quat. Int.*, *357*, 110-124.

Bierman, P., Marsella, K., Patterson, C., Davis, P. & Caffee, M. (1999). Mid-Pleistocene cosmogenic minimum age limits for the pre-Wisconsinan glacial surfaces in southwestern Minnesota and southern Baffin Island: a multiple-nuclide approach. *Geomorphology*, *27*, 25-39.

Bird, M. I., Fifield, L. K., Santos, G. M., Beaumont, P. B., Zhou, Y., di Tada, M. L. & Hausladen, P. A. (2003). Radiocarbon dating from 40 to 60 ka BP at Border Cave, South Africa. *Quat. Sci. Rev.*, *22*, 943-947.

Bøe, A. G., Murray, A. S. & Olaf Dahl, S. (2007). Resetting of sediments mobilised by the LGM ice sheet in southern Norway. *Quat. Geochronol.*, *2*, 222-228.

Bowman, S. (1990). *Radiocarbon Dating*. London: British Museum.

Braconnot, P., Harrison, S. P., Kageyama, M., Patrick, J., Bartlein, P. J., Masson-Delmotte, V., Abe-Ouchi, A., Otto-Bliesner, B. & Zhao, Y. (2012). Evaluation of climate models using palaeoclimatic data. *Nat. Clim. Change*, *2*, 417-424.

Brauer, A., Anreas, C. & Negendank, J. F. W. (1999). Lateglacial calendar year chronology based on annually laminated sediments from Lake Meerfelder Maar, Germany. *Quat. Int.*, *61*, 17-25.

Breckenridge, A. (2007). The Lake Superior varve stratigraphy and implications for eastern Lake Agassiz outflow from 10,700 to 8900 cal ybp (9.5-8.0 $^{14}$C ka.). *Palaeogeogr. Palaeoclimatol. Palaeoecol.*, *249*, 45-61.

Breckenridge, A., Lowell, T. V., Stroup, J. S. & Evans, G. (2012). A review and analysis of varve thickness records from glacial Lake Ojibway (Ontario and Quebec, Canada). *Quat. Int.*, *260*, 43-54.

Briner, J. (1998). Using inherited cosmogenic Cl-36 to constrain glacial erosion rates of the Cordilleran ice-sheet. *Geology*, *26*, 3-6.

Broecker, W. S. & Denton, G. H. (1990). What drives glacial cycles? *Sci. Am.*, *262*, 42-50.

Broecker, W. S., Kennet, J. P., Flower, B. P., Teller, J. T., Trumbore, S., Bonani, G. & Wolfi, W. (1989). Routing of meltwater from the Laurentide Ice sheet during the Younger Dryas cold episode. *Nature*, *341*, 318-321.

Bruaer, A., Hajdas, I., Blockley, S. P. E., Bronk Ramsey, C., Christil, M., Ivy-Ochs, S., Moseley, G. E., Nowaczyk, N. N., Rasmussen, S. O., Roberts, H. M., Spötl, C., Staff, R. A. & Svensson, A. (2014). The importance of independent chronology in integrating records of past climate change for the 60-8 ka INTIMATE time interval. *Quat. Sci. Rev.*, *106*, 47-66.

Buntzin, et al. (2016). Marine radiocarbon reservoir age simulations for the past 50000 years. EGU General Assembly, Vienna, Austria.

Buylaert, J. P., Thiel, C., Murray, A. S., Vandenberghe, D. A. G., Yi, S. & Lu, H. (2011). IRSL and Post-IR IRSL residual doses recorded in modern dust samples from the Chinese loess plateau. *Geochronometria*, *38*, 432-440.

Buylaert, J. P., Yeo, E. Y., Thiel, C., Yi, S., Stevens, T., Thompson, W., Frechen, M., Murray, A. & Lu, H. (2015). A detailed post-IR IRSL chronology for the last interglacial soil at the Jingbian loess site (northern China). *Quat. Geochronol.*, *30*, 194-199.

Cane, M. A. & Molnar, P. (2001). Closing of the Indonesian seaway as precursor to east African aridification around 3-4 million years ago. *Nature*, *411*, 157-162.

Cauquoin, A., Raisbeck, G., Jouzel, J. & Paillard, D. (2014). Use of $^{10}$Be to predict atmospheric $^{14}$C variations during the Laschamp Excursion: High sensitivity to cosmogenic isotope production calculations. *Radiocarbon*, *56*, 67-82.

Cheng, H., Edwards, R. L., Broecker, W. S., Denton, G. H., Kong, X., Wang, Y., Zhang, R. & Wang, X. (2009). Ice age terminations. *Science*, *326*, 248-252.

Clark, P. U., Shakun, J. D., Baker, P. A., Bartlein, P. J., Brewer, S., Brook, E., Carlson, A. E., Cheng, H., Kaufman, D. S., Zhengyu, L., Marchitto, T. M., Mic, A. C., Morril, C., Otto-Bliesner, B. L., Pahnke, K., Russel, J. M., Whitlock, C., Adkins, J. F., Blois, J. L., Clark, L., Colman, S. M., Curry, W. B., Flower, B. P., He, F., Johnson, T. C., Lynch-Stieglitz, J., Markgraf, V., McManus J., Mitrovica, J. X., Moreno, P. I.. & Williams, J. W. (2012). Global climate evolution during the last deglaciation. *PNAS*, *109*(19) doi: 10.1073/pnas.1116619109.

Clayton, L. & Moran, S. R. (1982). Chronology of Late Wisconsinan glaciation in middle North America. *Quat. Sci. Rev.*, *1*, 55-82.

Coates, A. G. & Stallard, R. F. (2013). How old is the Isthmus of Panama? *Bull. Mar. Sci.*, *89*, 801-813.

Colgan, P. M., Bierman, P. R., Mickelson, D. M. & Caffee, M. (2002). Variation in glacial erosion near the southern margin of the Laurentide Ice Sheet, south-central Wisconsin, USA: Implications for cosmogenic dating of glacial terrains. *Geol. Soc. Am. Bull.*, *114*, 1581-1591.

Cooke, R., Warren, A. & Goudie, A. (1993). *Desert Geomorphology*. UCL Press: London.

Cuffrey, K. M., Clow, G. D., Steig, E. J., Buizert, C., Fudge, T. J., Koutnic, M., Waddington, E. D., Alley, R. B. & Severinghaus, J. P. (2016). Deglacial temperature history of west Antarctica. *PNAS*, *113* (50) 14249-14254.

Dahl-Jensen, D., Gundestrup, N., Gogineni, S. P. & Miller, H. (2003). Basal melt at NorthGrip modeled from boreholes, ice-core and radio-echo sounder observations. *Ann. Glaciol.*, *37*, 207-212.

Damon, P. E. & Peristykh, A. N. (2000). Radiocarbon calibration and application to geophysics, solar physics and astrophysics. *Radiocarbon*, *42*, 137-150.

Darvill, C. M. (2013). *Cosmogenic Nuclide Analysis.* In Clarke, L., editor. Geomorphological Techniques. London: British Society for Geomorphology.

Davis, P., Bierman, P., Marsella, K., Caffee, M. & Southon, J. (1999). Cosmogenic analysis of glacial terrains in the eastern Canadian arctic: a test for inherited nuclides and the effectiveness of glacial erosion. *Annals of Glaciology*, *28*, 181-188.

De Geer, G. (1912). A geochronology of the last 12,000 years. XIth *International Geological Congress*, Stockholm, *1*, 241-253,

De Vries, H. (1958). Variations in concentration of radiocarbon with time and location on earth. *KNAW*, *Proceedings*, *B61*, 94-102.

Dean, M. J., Kemp, A. E. S., Bull, D., Pike, J., Pettersen, G. & Zolitschka, B. (1999). Taking varves to bits: scanning microscopy in the study of laminated sediments and varves. *J. Paleolimnol.*, *22*, 121-136.

Denton, G. H., Anderson, R. F., Toggweiler, J. R., Edwards, R. L., Schaefer, J. M. & Putman, A. E. (2010). The last glacial termination. *Science*, *238*, 1652-1656.

Duller, G. A. T. (2006). Single grain optical dating of glacigenic deposits. *Quat. Geochronol.*, *1*, 296-304.

Dunai, T. (2010). Cosmogenic nuclides: Principles, concepts and applications in the earth surface sciences. Cambridge: Cambridge University Press,

Dunai, T. J. & Lifton, N. A. (2014). The nuts and bolts of cosmogenic nuclide production. *Elements*, *10*, 347-350.

Dyke, A. S. (2004). An outline of North American deglaciation with emphasis on central and northern Canada. In Ehlers, J., Gibbard, P.l., editors. Quaternary Glaciation–Extent and chronology, Part II. *Developments in Quaternary Science*, Volume *26*. The Netherlands: Elsevier.

Dyke, A. S. (2005). Late Quaternary vegetation history of northern North America based on pollen, macrofossil, and faunal remains. *Géogr. Phys. Quat.*, 592-3, 211-262.

Dyke, A. S., Andrews, J. T., Clark, P. U., England, J. H., Miller, G. H., Shaw, J. & Veillette, J. J. (2002). The Laurentide and Innuitian ice sheets during the Last Glacial Maximum. *Quat. Sci. Rev.*, *21*, 9-31.

Dyke, A. S., Andrews, J. T., Clark, P. U., England, J. H., Miller, G. H., Shaw, J. & Veillete J. J. (2001). Radiocarbon dates pertinent to defining the Last Glacial Maximum for the Laurentide and Inuitian ice sheets. Ottawa: Geological Survey of Canada Open File, *412*, pp. 50.

Dyke, A. S., Moore, A. J. & Robertson, L. (2003). *Deglaciation of North America*. Geological Survey of Canada, Open File, *1574*, 32 maps, CD-ROM and paper sheets.

Fabel, D., Harbor, J., Dahms, D., James, A., Elmore, D., Horn, L., Daley, K. & Steele, C. (2004). Spatial patterns of glacial erosion at a valley scale derived from terrestrial cosmogenic $^{10}$Be and $^{26}$Al concentrations in rock. *Ann. Ass. Am. Geog.*, *92*, 241-255.

Fabel, D. J., Stone, J., Fifield, L. K. & Cresswell, R. G. (1997). Degaciation of the Vestford Hills, East Antarctica: Preliminary evidence from exposure dating of three subglacial erratics. In Ricci, C. A., editor. The Antarctic region: Geological Evoluation and Processes. Siena: Terra Antarctica.

Fisher, T. G., Waterson, N., Lowell, T. V. & Hajdas, I. (2009). Deglaciation ages and meltwater routing in the Fort McMurray region. Northeastern Saskatchewan, Canada. *Quat. Sci. Rev.*, *28*, 1608-1624.

Fleming, S. J. (1970). Thermoluminescent dating: refinement of the quartz inclusion method. *Archaeometry*, *12*, 133-45.

Francus, P., Ridge, J. C. & Johnson, M. D. (2013). The rise of varves. *GFF*, *135*, 229-230.

Frouin, M., Huot, S., Kreutzer, S., Lahaye, C., Lamothe, M., Philippe, A. & Mercier, N. (2017). An improved radiofluorescence single-aliquot regenerative dose protocol for K-feldspars. *Quat. Geochronol.,* 38, 13-24.

Fuchs, M. & Owen, L. A. (2008). Luminescence dating of glacial and associated sediment: review, recommendations and future directions. *Boreas*, *31*, 636-659.

Gaar, D., Lowick, S. E. & Preusser, F. (2013). Performance of different luminescence approaches for the dating of known-age glaciofluvial deposits from northern Switzerland. *Geochronometria*, *41*, 65-80.

Garzione, C. N. (2008). Surface uplift of Tibet and Cenozoic global cooling. *Geology*, *36*, 1003-1004.

Gibbard, P. & Head, M. J. (2009). The definition of the Quaternary System/Era and the Pleistocene Series/Epoch. *Quaternaire*, *20*, 125-133.

Gibbard, P. L., Head, M. J. & Walker, M. J. C. The Subcommission on Quaternary Stratigraphy, (2010). Formal ratification of the Pleistocence Series/Epoch with a base at 2.58 Ma. *J. Quat. Sci.*, *25*, 96-102.

Godfrey-Smith, D. I., Huntley, D. J. & Chen, W. H. (1988). Optical dating studies of quartz and feldspar sediment extracts. *Quat. Sci. Rev.*, *7*, 373-380.

Gosse, J. C. (2005). The contributions of cosmogenic nuclides to unraveling alpine paleo-glacier histories. In: Huber, U. M., Bugman, H. K. M., Reasoner, M. A., editor. Global change and Mountain Regions. A state of knowledge overview, 39-50. Dordrecht: Springer.

Gosse, J. C. & Phillips, F. M. (2001). Terrestrial in situ cosmogenic nuclides: theory and application. *Quat. Sci. Rev.*, *20*, 1475-1560.

Gosse, J. C., Klein, J., Evenson, E. B., Lawn, B. & Middleton, R. (1995). Berrylium-10 dating of the duration and retreat of the last Pinedale glacial sequence. *Science*, *268*, 1329-1333.

Granger, D., Lifton, N. & Willenbring, J. (2013). A cosmic trip: 25 years of cosmogenic nuclides in geology. *Geol. Soc. Am. Bull.*, *125*, 1379-1402.

Guerin, G., Mercier, N. & Adamiec, G. (2011). Dose-rate conversion factors: update: *Ancient TL*, *29*, 5-8.

Guido, Z., Ward, D. & Anderson, R. (2007). Pacing the post-Last Glacial Maximum demise of the Animas Valley glacier and the San Juan Mountain ice cap. Colorado. *Geology*, *35*, 739-742.

Halfen, A. F., Lancaster, N. & Wolfe, S. (2015). Interpretations and common challenges of Aeolian records from North American dune fields. *Quat. Int.*, *410*, 75-95.

Haltia-Hovi, E., Saarinen, T. & Kukkonen, M. (2007). A 2000-year record of solar forcing on varved sediment in eastern Finland. *Quat. Sci. Rev.*, *26*, 678-689.

Harden, J. W. (1982). A quantitative index of soil development from field descriptions: examples from a chronosequence in central California. *Geoderma*, *28*, 1-28.

Harkness, D. D. (1979). Radiocarbon dates from Antarctica. *Brit. Antarct. Surv. Bull.*, *47*, 43-59.

Hartman, D. M. D. & Clague, J. J. (2008). Quaternary stratigraphy and glacial history of the Peace River valley, northeast British Columbia. *Can. J. Earth Sci.*, *45*, 549-564.

Haug, G. H., Ganopolski, A., Sigman, D. M., Rosell-Mele, A., Swann, G. E. A., Tidemann, R., Jaccard, S. L., Bollmann, J., Maslin, M. A., Leng, M. J. & Eglington, G. North Pacific seasonality and the glaciation of North America 2.7 million years ago. *Nature*, *433*, 821-825.

Haug, G. H. & Tiedemann, R. (1998). Effect of the formation of the Isthmus of Panama on Atlantic Ocean thermohaline circulation. *Nature*, *393*, 673-676.

He, F., Shakun, J. D., Clark, P. U., Carlson, A. E., Liu, Z., Otto-Bliesner, B. L. & Kutzback, J. E. (2013). *Nature*, *494*, 81-85.

Head, M. J., Gibbard, P. L. & Slavador, A. (2008). The Quaternary: its character and definition. *Episodes*, *31*, 234-238.

Hilgen, F., Brinkhuis, H. & Zachariase, W. J. (2006). Unit stratotypes for global stages: the Neogen perspective. *Earth Sci. Rev.*, *74*, 113-125.

Hill, C. (2006). Geological framework and glaciation of the central area. In Ubelaker, D. H., editor. Handbook of North American Indians. Environments, Origins and Population. Washington D.C.: Smithsonian Institution, pp. 67-80.

Hugget, R. J. (1998). Soil chronosequences, soil development, and soil evolution: a critical review. *Catena*, *32*, 155-172.

Hughes, A. L. C., Gyllencreutz, R., Lohne, O. S., Mangerud, J. & Svendsen, J. I. (2016). The last Eurasian ice sheets – a chronological database and time-slice reconstruction, DATED-1. *Boreas*, *45*, 1-45.

Huntley, D. J., Godfrey-Smith, D. I. & Thewalt, M. L. W. (1985). Optical dating of sediments, *Nature*, *313*, 105-107.

Ivy-Ochs, S. & Kober, F. (2008). Surface exposure dating with cosmogenic nuclides. *Quaternary Science Journal*, *57*, 179-209.

Jackson, J. B. C. & O'Dea, A. (2013). Timing of the oceanographic and biological isolation of the Caribbean Sea from the Tropical Eastern Pacific Ocean. *Bull. Mar. Sci.*, *89*, 779-800.

Jackson, Jr. L., Phillips, F. M., Shimamura, K. & Little, E. C. (1997). Cosmogenic$^{36}$Cl dating of the Foothills erratics train, Alberta, Canada. *Geology*, *25*, 195-198.

Jain, M., Murray, A. S. & Bøtter-Jensen, L. (2003). Characterisation of blue-light stimulated luminescence components in different quartz samples: implications for dose measurement. *Radiat. Meas.*, *37*, 441-449.

Johnson, M. D., Kylander, M. E., Casserstedt, L., Wiborgh, H. & Björck, S. (2012). Varved glaciomarine clay in central Sweden before and after the Baltic Ice lake drainage: a further clue to the drainage events at Mt Billingen. *GFF*, *135*, 293-307.

Johnston, W. A. (1933). Quaternary geology of North America in relation to the migration of man. In Jennes, D., editor. The American Aborigines. University of Toronto Press, Toronto, pp. 11-45.

Jouzel, J., Masson-Delmotte, V., Cattani, O., Dreyfus, G., Falourd, S., Hoffman, G., Minster, B., Nouet, J., Barnola, J. M., Chappellaz, J., Fischer, H., Gallet, J. C., Johnsen, S., Leunebeger, M., Loulergue, L., Luethi, D., Oerter, H., Parrenin, F., Raisbeck, G., Raynaud, D., Schilt, A., Schwander J., Selmo, E., Souchez, R., Spahni, R., Stauffer, B., Steffensen, J. P., Stenni, B., Stocker, T. F., Tison, J. l., Werner, M. & Wolff, E. W. (2007). Orbital and millennial Antarctic climate variability over the past 800,000 years. *Science*, *317*, 793-796.

Keigwin, L. (1982). Isotopic paleoceanography of the Caribbean and east Pacific: Role of Panama uplift in late Neogene time. *Science*, *217*, 350-353.

Keigwin, L. D. (1978). Pliocene closing of the Isthmus of Panama, based on biostratigraphic evidence from nearby Pacific Ocean and Caribbean Sea cores. *Geology*, *6*, 630-634.

Kelly, M. A., Ivy-Ochs, S., Kubik, P. W., von Blankenburg, F. & Schlüchter, C. (2006). Chronology of deglaciation based on $^{10}$Be dates of glacial erosional features in the Grimsel Pass region, central Swiss Alps. *Boreas*, *35*, 634-643.

King, G. E., Robinson, R. A. J. & Finch, A. A. (2013). Apparent OSL ages of modern deposits from Fåbergstølsdalen, Norway: implications for sampling glacial sediments. *J. Quat. Sci.*, *28*, 673-682.

Klasen, N., Fiebig, M., Preusser, F. & Radtke, U. (2006). Luminescence properties of glaciofluvial sediments from the Bavarian Alpine Foreland. *Radiat. Meas.*, *41*, 866-870.

Lai, D. (1991). Cosmic ray labeling of erosion surfaces – In situ nuclide production rates and erosion models. *Earth Planet. Sci. Lett.*, *104*, 424-439.

Lal, D. & Peters, B. (1967). Cosmic ray produced radioactivity on earth. In Flugg S., editor. *Handbook of Physics*, vol 46/2. Berlin: Springer pp. 551-612.

Lamothe, M. & Auclair, M. (1999). A solution to anomalous fading and age shortfalls in optical dating of feldspar minerals. *Earth Planet. Sci. Lett.*, *171*, 319-323.

Larocque-Tobler, I., Filipiak, J., Tylmann, W. & Bonk, A. (2015). Comparison between chironomid-inferred mean-August temperature from varved Żabińskie (Poland) and instrumental data since 1896 AD. *Quat. Sci. Rev.*, *111*, 35-50.

Lasberg, K. & Kalm, V. (2013). Chronology of late Weichselian glaciation in the western part of the East European Plain. *Boreas*, *42*, 995-1007.

Libby, W. F. (1946). Atmospheric Helium three and radiocarbon from cosmic radiation. *Phys. Rev.*, *69*, 671-672.

Libby, W. F. (1952). *Radiocarbon dating*. Chicago: University of Chicago Press.

Libby, W. F., Anderson, E. C. & Arnold, J. R. (1949). Age determination by radiocarbon content: world-wide assay of natural radiocarbon. *Science*, *109*, 227-228.

Licciardi, J. M., Clark, P. U., Brook, E. J., Elmore, D. & Sharma, P. (2004). Variable responses of western U.S. glaciers during the last deglaciation. *Geology*, *32*, 81-84.

Lifton, N., Bieber, J., Clem, J., Duldi, M., Evenson, P., Humble, J. & Pyle, R. (2005). Addressing solar modulation and long-term uncertainties in scaling secondary cosmic rays for in situ cosmogenic nuclide applications. *Earth Planet. Sci. Lett.*, *239*, 140-161.

Lifton, N., Sato, T. & Dunai, T. J. (2014). Scaling in situ cosmogenic nuclide production rates using analytical approximations to atmospheric cosmic-ray fluxes. *Earth Plant. Sci. Lett.*, *386*, 149-160.

Logie, M. (1982). Influence of roughness elements and soil moisture on the resistance of sand to wind erosion. *Catena Suppl.*, *1*, 161-173.

Lowe, J. & Walker, M. (2015). *Reconstructing Quaternary environments* (3rd ed). Oxford, UK: Routledge.

Lucke, A., Schleser, S. G., Zolitschka, B. & Negendank, J. F. W. (2003). Lateglacial and Holocene organic carbon isotope record of lacustrine palaeoproductivity and climate change derived from varved lake sediments of Lake Holzmaar. Germany. *Quat. Sci. Rev.*, *22*, 569-580.

Lukas, S., Spencer, J. Q., Robinson, R. A. J. & Benn, D. J. (2007). Problems associated with luminescence dating of Late Quaternary glacial sediments in the NW Scottish Highlands. *Quat. Geochronol., 2*, 243-248.

Madsen, D. B. (2015). A framework for the initial occupation of the Americas. *PaleoAmerica, 1*(3), 217-249.

Mandryk, C. A. S., Josenhans, H., Fedje, D. W. & Mathewes, R. W. (2001). Late Quaternary paleoenvironments of northwestern North America: implications for inland versus coastal migration routes. *Quat. Sci. Rev., 20*, 301-314.

Martin-Puertas, C., Tjallingii, R., Bloemsma, M. & Brauer, A. (2017). Varved sediment responses to early Holocene climate and environmental changes in Lake Meerfleder Maar (Germany) obtained from multivariate analyses of micro X-ray fluorescence core scanning data. *J. Quat. Sci., 32*, 427-436.

Masarik, J. & Reedy, R. (1995). Terrestrial cosmogenic-nuclide production systematics calculated from numerical simulations. *Earth Planet. Sci. Lett., 136*, 381-395.

Masarik, J. & Beer, J. (1999). Simulation of particle fluxes and cosmogenic nuclide production in the Earth's atmosphere. *J. Geophys. Res. D, 104*, 12099-12111.

Masarik, J. & Wieler, R. (2003). Production rates of cosmogenic nuclides in boulders. *Earth Planet. Sci. Lett., 216*, 201-208.

Masson-Delmonte, V., Stenni, B., Pol, K., Braconnot, P., Cattani, O., Falourd, S., Kageyama, M., Jouzel, J., Landais, A., Minster, B., Barnola, J. M., Chappellaz, J., Krinner, G., Johnsen, S., Röthlisberger, R., Hansen, J., Mikolajewicz, U. & Otto-Bliesner, B. (2010). Epica Dome C record of glacial and interglacial intensities. *Quat. Sci. Rev., 29*, 113-128.

McCabe, A. M. (1996). Dating and rhythmicity from the last deglacial cycle in the British Isles. *J. Geol. Soc. London, 153*, 499-502.

McCabe, A. M., Clark, P. U. & Clark, J. (2005). AMS $^{14}C$ dating of deglacial events in the Irish Sea Basin and other sectors of the British-Irish ice sheet. *Quat. Sci. Rev., 24*, 1673-1670.

McKeever, S. W. S., Moscovitch, M. & Townsend, P. D. (1995). *Thermoluminescence dosimetry: Properties and Uses*. Kent: Nuclear Technology Publishing.

Mickelson, D. M. & Colgan, P. M. (2003). The southern Laurentide Ice Sheet. *Development in Quaternary Science*, *1*, 1-16.

Molnar, P. (2008). Closing of the Central American Seaway and the Ice Age: A critical review. *Paleoceanography*, *23*, PA2201, doi:10.1029/2007PA001574.

Montes, C., Cardona, A., Jaramillo, C., Pardo, A., Silva, J. C., Valencia, V., Ayala, C., Pérez-Angel, L. C., Rodriguez-Parra, A., Ramirez, V. & Niño, H. (2015). Middle Miocene closure of the Central American Seaway. *Science*, *348*, 226-229.

Munyikwa, K., Feathers, J. K., Rittenour, T. & Shrimpton, H. K. (2011a). Constraining the Late Wisconsinan retreat of the Laurentide Ice Sheet from western Canada using luminescence ages from postglacial eolian dunes. *Quat. Geochronol.*, *6*, 407-422.

Munyikwa, K., Telfer, M., Baker, I. & Knight, C. (2011b). Core drilling of Quaternary sediments for luminescence dating using the Dormer Drillmite™. *Ancient TL*, *29*, 15-22.

Munyikwa, K., Rittenour, T. M. & Feathers, J. K. (2017). Temporal constraints for the Late Winsconsinan deglaciation of western Canada using eolian dune luminescence chronologies from Alberta. *Palaeogeog. Palaeoclim. Palaeoecol.*, *470*, 147-165.

Murray, A., Buylaert, J. P., Henriksen, M., Svendsen, J. I. & Mangerud, J., (2008). Testing the reliability of quartz OSL ages beyond the Eemian. *Radiat. Meas.*, *43*, 776-780.

Murray, A. S. & Roberts, R. G. (1998). Measurement of the equivalent dose in quartz using a regenerative–dose single aliquot protocol. *Radiat. Meas.*, *29*, 503-515.

Murray, A. S. & Wintle, A. G. (2000). Luminescence dating of quartz using an improved single-aliquot regenerative-dose protocol. *Radiat. Meas.*, *32*, 57-73.

Murray, A. S. & Wintle, A. G. (2003). The single aliquot regenerative dose protocol: potential improvements in reliability. *Radiat. Meas.*, *37*, 377-381.

Nakagawa, T., Gotanda, K., Haraguchi, T., Danhara, T., Yonenobu, H., Bruaer, A., Yokohama, Y., Tada, R., Takemura, K., Staff, R. A., Payne, R., Brock Ramsey, C., Bryant, C., Brock, F., Schlolaut, G., Marchall, M., Tarasov, P., Lamb, H. & Suigetsu Project Members, (2012). SG06, a fully continuous and varved sediment core from Lake Suigetsu, Japan: stratigraphy and potential for improving the radiocarbon calibration model and understanding of late Quaternary climate changes. *Quat. Sci. Rev.*, *36*, 164-176.

Nydal, R. (1966). Variation in C concentration in the atmosphere during the last several years. *Telus*, *XVIII*, (2) 271-279.

O. Dea, A., Lessios, H. A., Coates, A. G., Eytan, R. I., Restrepo-Moreno, S. A., Cione, A. L., Collins, L. S., de Queiroz, A., Farris, D. W., Norris, R. D., Stallard, R. F., Woodbourne, M. O., Aguilera, O., Aubry, M. P., Berggren, W. A., Budd, A. F., Cozzuol, M. A., Coppard, S. E., Duque-Caro, H., Finnegan, S., Gasparin, G. M., Grossman, E. L., Johnson, K. G., Keigwin, L. D., Knowlton, N., Leigh, E. G., Loenard-Pingel, J. S., Marko, P. B., Pyenson, N. D., Rachello-Dolmen, P. G., Soibelzon, E., Soibelzon, L., Todd, J. A., Vermeij, G. J. & Jackson, J. B. C. (2016). Formation of the Isthmus of Panama. *Science Advances*, Vol. *2*, no. 8, e1600883 doi: 10.1126/sciadv.1600883.

Ojala, A. E. K. & Alenius, T. (2005). 10 000 years of interannual sedimentation recorded in the Lake Nautajarvi (Finland) clastic-organic varves. *Palaeogeo. Palaeoclim. Palaeoecol.*, *219*, 285-302.

Olley, J. M., Roberts, R. G., Yoshida, H. & Bowler, J. M. (2006). Single-grain optical dating of grave infill associated with human burials at Lake Mungo, Australia. *Quat. Sci. Rev.*, *25*, 2469-74.

Paillard, D. & Parrenin, F. (2004). The Antarctic ice sheet and the triggering of deglaciations. *Earth Planet. Sci. Lett.*, *227*, 263-271.

Pawley, S. M., Bailey, R. M., Rose, J., Moorlock, B. S. P., Hamblin, R. J. O., Booth, S. J. & Lee, J. R. (2008). Age limits on Middle Pleistocene glacial sediments from OSL dating, north Norfolk, UK. *Quat. Sci. Rev.*, *27*, 1363-1377.

Phillips, F. M., Zreda, M. G., Smith, S. S., Elmore, D., Kubik, P. W. & Sharma, P. (1990). Cosmogenic chlorine-36 chronology for glacial deposits at Bloody Canyon, eastern Sierra-Nevada. *Science*, *248*, 1529-1532.

Phillipsen, B. (2013). The freshwater reservoir effect in radiocarbon dating. *Herit. Sci.*, *1*, 24. doi:10.1186/2050-7445-1-24.

Pietsch, T. J., Olley, J. M. & Nanson, G. C. (2008). Fluvial transport as a natural luminescence sensitizer of quartz. *Quat. Geochronol.*, *3*, 365-376.

Prescott, J. R. & Hutton, J. T. (1994). Cosmic ray contribution to dose rates for luminescence and ESR dating. *Radiat. Meas.*, *23*, 497-500.

Putkonen, J. & O'Neal, M. (2006). Degradation of unconsolidated Quaternary landforms in the western North America. *Geomorphology*, *75*, 408-419.

Putkonen, J. & Swanson, T. (2003). Accuracy of cosmogenic ages for moraines. *Quat. Res.*, *59*, 255-261.

Raukas, A. & Stankowski, W. (2005). Influence of sedimentological composition on OSL dating of glaciofluvial deposits: examples from Estonia. *Geol. Quarterly*, *49*, 463-470.

Raukas, A., Stankowski, W. T. J., Zelčs, V. & Šinkunas, P. (2010). Chronology of the last deglaciation in the south-eastern Baltic region on the basis of recent OSL dates. *Geochronometria*, *36*, 47-54.

Ravelo, A. C., Anderson, D. H., Lyle, M., Olivarez Lyle, A. & Wara, M. W. (2004). Regional climate shifts caused by gradual global cooling in the Pliocene epoch. *Nature*, *429*, 263-267.

Raymo, M. E. & Ruddiman, W. F. (1992). Tectonic forcing of late Cenozoic climate. *Nature*, *359*, 117-122.

Rees-Jones, J. (1995). Optical dating of young sediments using fine grain quartz. *Ancient* TL, *13*, 9-14.

Reimer P. J. & Reimer, R. W. (2001). A marine reservoir correction database and on-line interface. *Radiocarbon*, *43*, 461-463.

Reimer, P. J., Baillie, M. G. L., Bard, E., Bayliss, A., Beck, J. W., Blackwell, P. G., Bronk Ramsey, C., Buck, C. E., Burr, G. S., Edwards, R. L., Friedrich, M., Grootes, P. M., Guilderson, T. P., Hajdas, I., Heaton, T. J., Hogg, A. G., Hughen, K. A., Kaiser, K. F., Kromer, B., McCormac, F. G., Manning, S. W., Reimer, R. W., Richard, D. A., Southon, J. R., Talamo, S., Turney, C. S. M., van der Plicht, J. & Weyhenmeyer, C. E. (2009). IntCal09 and Marine09 radiocarbon age calibration curves, 0-50,000 years cal BP. *Radiocarbon*, *51*, 1111-1150.

Reimer, P. J., Baillie, M. G. L., Bard, E., Bayliss, A., Beck, J. W., Bertrand, C. J. H., Blackwell, P. G., Buck, C. E., Burr, G. S., Cutler, K. B., Damon, P. E., Edwards, R. L., Fairbanks, R. G., Freidrich, M., Guilderson, T. P., Hogg, A. G., Hughen, K. A., Kromer, B., McCormac, G., Manning S., Bronk Ramsey, C., Riemer, R. W., Remmele, S., Southon J. R., Stuiver M., Talamo, S., Taylor, F. W., van der Plicht, J. & Weyhenmeyer C. E. (2004). IntCal04 Terrestrial radiocarbon age calibration, 0-26 cal kyr BP. *Radiocarbon*, *46*, 1029-1058.

Reimer, P. J., Bard, E., Bayliss, A., Beck, J. W., Blackwell, P. G., Bronk Ramsey, C., Buck, C. E., Cheng, H., Edwards, R. L., Friedrich, M., Grootes, P. M., Guilderson, T. P., Haflidason, H., Hajdas, I., Hatte, C., Heaton, T. J., Hoffmann, D. L., Hogg, A. G., Hughen, K. A., Kaiser, K. F., Kromer, B., Manning, S. W., Niu, M., Reimer, R. W., Richards, D. A., Scott, E. M., Southon, J. R., Staff, R. A., Turney, C. S. M. & van der Plicht, J. (2013). IntCal13 and Marine13 radiocarbon age calibration curves 0-50,000 years cal BP. *Radiocarbon*, *55*, 1869-1887.

Reimer, P. J., McCormac, F. G., Moore, J., McCormick, F. & Murra, E. V. (2002). Marine radiocarbon reservoir corrections for the mid- to late Holocene in the eastern subpolar North Atlantic. *Holocene*, *12*, 129-135.

Rhodes E. J. & Bailey R. M. (1997). The effect of thermal transfer on the zeroing of the luminescence of quartz from recent glaciofluvial sediments. *Quat. Geochronol.*, *16*, 291-298.

Rhodes, E. J. (2000). Observations of thermal transfer OSL signals in glacigenic quartz. *Radiat. Meas.*, *32*, 595-602.

Rhodes, E. J. (2011). Luminescence dating of sediment over the past 200,000 years. *Annu. Rev. Earth Planet. Sciences*, *39*, 461-488.

Richards, B. M. W. (2000). Luminescence dating of quaternary sediment in the Himalaya and High Asia: a practical guide to its use and limitations for constraining the timing of glaciation. *Quat. Int.*, *65*, 49-61.

Richter, D., Richter, A. & Dornich, K. (2011). Lexsyg - A new system for luminescence research. *Geochronometria*, *20*(4), 220-228.

Ridge, J., Balco, G., Bayless, R. L., Beck, C. C., Carter, L. B., Dean, J. L., Voytek, E. B. & Wei, J. H. (2012). The new North American varve chronology: a precise record of southeastern Laurentide Ice Sheet deglaciation and climate, 18.2-1.5 kyr BP, and correlations with Greenland ice core records. *Am. J. Sci.*, *312*, 685-722.

Ridge, J. C. (2003). The Last Deglaciation of the Northeastern United States. A Combined Varve, Paleomagnetic and Calibrates $^{14}C$ Chronology. In Hart J. P., Cremeens, D. L., editors. Geoarcheology of Landscapes in the Glaciated Northeast U. S. New York State Museum Bulletin 497, 15-45.

Ridge, J. C. (2004). The Quaternary glaciation of western New England with correlations to surrounding areas. In Ehlers, J., Gibbard, P.l., editors. Quaternary Glaciations – Extent and Chronology, Part II: North America. Amsterdam, Elsevier, Developments in Quaternary Science 2b, 163-193.

Rittenour, T. M., Brigham-Grett, J. & Mann, M. E. (2000). El Nino-like climate teleconnections in New England during the Late Pleistocene. *Science*, *288*, 1039-1042.

Rupper, S., Roe, G. & Gillespie, A. (2009). Spatial patterns of Holocene glacier advance and retreat in Central Asia. *Quat. Res.*, *72*, 337-346.

Saarni, S., Saarinen, T. & Lensu, A. (2015). Organic lacustrine sediment varves as indicators of past precipitation changes: a 3,000-year climate record from central Finland. *J. Paleolimnol.*, *53*, 401-413.

Sanderson, D. C. W., Bishop, P., Stark, M., Alexander, S. & Penny, D. (2007). Luminescence dating of canal sediments from Angor Borei, Mekong Delta, southern Cambodia. *Quat. Geochron.*, *2*, 322-329.

Sarnthein, M., Bartoli, G., Prange, M., Schmittner, A., Schneider, B., Weinelt, M., Andersen, N. & Garbe-Schonberg, D. (2009). Mid-Pliocene shift is ocean overturning circulation and the onset of Quaternary-style climates. *Clim. Past*, *5*, 269-283.

Sauerbrey, M. A., Juschus, O., Gebhardt, A. C., Wennrich, V., Nowaczyk, N. R. & Melles, M. (2013). Mass movement deposits in the 3.6 Ma sediment record of Lake El'gygytgyn, Far East Russian Arctic. *Clim. Past*, *9*, 1949-1967.

Schaetzl, R. J., Mikesell, L. R. & Velbel, M. A. (2006). Soil characteristics related to weathering and pedogenesis across a geomorphic surface of uniform age in Michigan. *Phys. Geogr.*, *27*, 170-188.

Schlolaut, G., Brauer, A., Marshall, M. H., Nakagawa, T., Saff, R. A., Bronk Ramsey, C., Lamb, H. F., Bryant, C. L., Naumann, R., Dulski, P., Brock, F., Yokohama, Y., Tada, R., Haraguchi, T. & Suigetsu, project members, (2014). Event layers in the Japanese Lake Suigetsu 'SG06' sediment core description, interpretation and climatic implications. *Quat. Sci. Rev.*, *83*, 157-170.

Severinghaus, J. P. (2009). Southern see-saw seen. *Nature*, *457*, 1093-1094.

Shackleton, N. J., Berger, A. & Peltier, W. R. (1990). An alternative astronomical calibration of the lower Pleistocene timescale based on ODP Site 677. *Trans. R. Soc. Edinburgh Earth Sci.*, *81*, 251-261.

Shackleton, N. J., Fairbanks, R. G., Chiu, T. C. & Parrenin, F. (2004). Absolute calibration of the Greenland time scale: implications for Antarctic time scales and for $\Delta\ ^{14}C$. *Quat. Sci. Rev.*, *23*, 1513-1522.

Shackleton, N. J. & Opdyke, N. D. (1973). Oxygen isotopes and palaeomagnetic stratigraphy of equatorial Pacific core V28-238: oxygen isotope temperatures and ice volume on $10^5$ and $10^6$ year scale. *Quat. Res.*, *3*, 39-55.

Shakun, J. D., Clark, P. U., He, F., Marcott, S. A., Mix, A. C., Liu, Z., Otto-Bliesner, B., Schmittner, A. & Bard, E. (2012). Global warming preceded by increasing carbon dioxide concentrations during the last deglaciation. *Nature*, *484*, 49–54.

Sigman, D. M., de Boer, A. M. & Haug, G. H. (2007). Antarctic stratification, atmospheric water vapor, and Heinrich events: A hypothesis for late Pleistocene deglaciations. *Geophys. Monog. Series 173: Ocean Circulation: Mechanisms and Impacts*, *AGU*, pp. 335-349.

Singhvi, A. K. & Porat, N. (2008). Impact of luminescence dating on geomorphological and palaeoclimate research in drylands. *Boreas*, *37*, 537-558.

Słowiński, M., Zawiska, I., Ott, F., Noryskiewicz, A. M., Plessenn, B., Apolinarska, K., Rzodkiewicz, M., Michczyńska, D. J. Wulf, S., Skubała, P., Kordowski, J., Błaszkiewicz, M. & Brauer, A. (2017). Differential proxy responses to late Allerod and early Younger Dryas climatic change recorded in varved sediments of the Trzechowskie palaeolake in northern Poland. *Quat. Sci. Rev.*, *158*, 94-106.

Smith, G. A. & Pun, A. (2010). How does earth work: physical geology and the process of science (2nd edition). New York: Prentice Hall.

Smith, V. C., Staff, R. A., Bockely, S. P. E., Bronk Ramsey, C., Nakagawa, T., Mark, D. F., Takemura, K., Danhara, T. & Suigetsu, (2006). Project Members, (2013). Identification and correlation of visible tephras in the Lake Suigetsu SG06 sedimentary archive, Japan: chronostratigraphic markers from synchronising of east Asian/west Pacific palaeoclimatic records across the last 150 ka. *Quat. Sci. Rev.*, *67*, 121-137.

Spencer, J. Q. & Owen, L. A. (2004). Optically stimulated luminescence dating of Late Quaternary glaciogenic sediments in the upper Hunza valley: validating the timing of glaciation and assessing dating methods. *Quat. Sci. Rev.*, *23*, 175-191.

Staiger, J., Gosse, J., Torantica, R., Oglesby, B., Fastook, J. & Johnson, J. V. (2007). Atmospheric scaling of cosmogenic nuclide production: Climate effect. *J. Geophys. Res.*, *112*, B02205, doi:10.1029/ 2005 JB003811.

Stuiver, M. & Suess, H. E. (1966). On the relationship between radiocarbon dates and true sample ages. *Radiocarbon*, *8*, 534-540.

Stokes, S. (1992). Optical dating of young sediments using quartz. *Quat. Sci. Rev. (Quat. Geochronol.)*, *11*, 153-159.

Stone, J. R. & Stone, B. D. (2012). Varve records from glacial lakes in southern New England: constraints on the timing of Late-Wisconsinan deglaciation. *Geol. Soc. Am. NE Section 47th Annual Meeting, Abstracts with Programs*, Vol. *44*, No. 2 p 71.

Stott, L., Timmermann, A. & Thunell, R. (2007). Southern hemisphere and deep-sea warming led deglacial atmospheric $CO_2$ rise and tropical warming. *Science*, *318*, 435-438.

Stroeven, A. P., Hättestrand, C., Kleman, J., Heyman, J, Fabel, D., Fredin, O., Goodfellow, B. W., Harbor, J. M., Jansen, J. D., Olsen, L., Caffee, M. W., Fink, D., Lundqvist, J., Rosqvist, G. C., Stromberg, B. & Jansson, K. N. (2016). Deglaciation of Fennoscandia. *Quat. Sci. Rev.*, *147*, 91-121.

Stuiver, M., Bruzanius, T. F., Becker, B. & Kromer, B. (1991). Climatic solar, oceanic and geomagnetic influences on Late Glacial and Holocene atmospheric $^{14}C/^{12}C$ change. *Quat. Res.*, *35*, 1-24.

Stuiver, M., Heusser, C. J. & Yang, C. (1978). North American glacial history extended to 75,000 years ago. *Science*, *200*, 16-21.

Stuiver, M., Reimer, P. J., Bard, E., Beck, J. W., Burr, G. S., Hughen, K. A., Kromer, B., McCormac, G., van der Plicht, M. & Spurk, M. (1998). IntCal98 radiocarbon age calibration, 24,000-0 cal BP. *Radiocarbon*, *40*, 1041-1084.

Svensson, A., Anderson, K. K., Bigler, M., Clausen, H. B., Dahl-Jensen, D., Davies, S. M., Johnsen, S. J., Muscheler, R., Parrenin, F., Rasmussen, S. O., Röthlisberger, R., Seierstad, I., Steffensen, J. P. & Vinther, B. M. (2008). A 60 000 year Greenland stratigraphic core chronology. *Clim. Past*, *4*, 47-57.

Tarasov, L. & Peltier, W. R. (2006). A calibrated deglacial drainage chronology for the North American continent: evidence of an arctic trigger for the Younger Dryas. *Quat. Sci. Rev.*, *25*, 659-688.

Taylor, R. E. (2016). Radiocarbon Dating: development of a Nobel Method. In Schuur, E. A. G., Druffel, E. R. M., Trumbore, S. E., editors. Radiocarbon and Climate Change: Mechanisms, Applications and Laboratory Techniques. Switzerland: Springer International.

Telfer, M. W., Wilson, P., Lord, T. C. & Vincent, P. J. (2009). New constraints of the age of the last ice sheet glaciations in NW England using optically stimulated luminescence dating. *J. Quat. Sci.*, *24*, 906-615.

Theodórsson, P. (1991). Gas proportional versus liquid scintillation counting radiometric versus AMS dating. *Radiocarbon*, *33*, 9-13.

Thornalley, D. J. R., McCave, N. I. & Elderfield, H. (2011). Tephra in deglacial ocean sediments south of Iceland: Stratigraphy, geochemistry and oceanic reservoir ages. *J. Quat. Sci.*, *26*, 190-198.

Thrasher I. M., Mauz B., Chiverrell R. C. & Lang, A. (2009). Luminescence dating of glaciofluvial deposits: a review. *Earth Sci. Rev.*, *97*, 133-146.

Tijlander, M., Ojala, A. E. K., Saarinen, T. & Snowball, I. (2002). Documentation of the physical properties of annually laminated (varved) sediments at a sub-annual to decadal resolution for environmental interpretation. *Quat. Int.*, *88*, 5-12.

Tijlander, M., Saarnisto, M., Ojala, E. E. K. & Saarinen, T. (2003). A 3000-year palaeoenvironmental record from annually laminated sediment of Lake Korttajärvi, central Finland. *Boreas*, *26*, 566-577.

Timar-Gabor, A., Constantin, D., Buylaert, J. P., Jain., M., Murray, A. S. & Wintle, A. G. (2015). Fundamental investigations of natural and laboratory generated SAR dose response curves for quartz OSL in the high dose range. *Radiat. Meas.*, *81*, 150-156.

Timmerman, A., Timm, O., Stott, L. & Menviel, L. (2009). The roles of $CO_2$ and orbital forcing in driving Southern Hemispheric temperature variations during the last 21 000 yr. *Journal of Climate*, *22*, 1626-1640.

Tsukamoto, S., Asahi, K., Watanabe, T. & Rink, W. J. (2002). Timing of past glaciations in Kanachenjunga Himal, Nepal by optically stimulated luminescence dating of tills. *Quat. Int.*, 97-98, 57-67.

Ullman, D. J., Carlson, A. E., Anslow, F. S., LeGrande, A. N. & Licciardi, J. M. (2015). Laurentide ice-sheet instability during the last deglaciation. *Nature Geosci.*, *8*, 534-537.

Vidic, N. J. & Lobnic, F. (1997). Rates of soil development of the chronosequences in the Ljubljana Basin, Slovenia. *Geoderma*, *76*, 35-64.

Walker, J. C. G., Hays, P. B. & Kasting, J. F. (1981). A negative feedback mechanism for the long-term stabilization of Earth's surface temperature. *J. Geophys. Res.*, *86*, 9776-9782.

Walker, M. (2012). *Dating Environmental Change and Constructing Chronologies*. In Matthews, J. A., Bartlein, P. J., Briffa, K. R., Dawson, A. G., Vernal, A. D., Denham, T., Fritz, S. C., Oldfield, F., editors. *The SAGE Handbook of Environmental Change*, Volume *1*. London: Sage Publications, pp. 92-115.

Walker, M., Johnsen, S., Rasmussen, S. O., Popp, T., Steffensen, J. P., Gibbard, P., Hoek, W., Lowe, J., Andrews, J., Björck, S., Cwynar, L. C., Hughen, K., Kershaw, P., Kromer, B., Litt, T., Lowe, D. J., Nakagawa, T., Newnham, R. & Schwander, J. (2009). Formal definition and dating of the GSSP (Global Stratotype Section and Point) for the base of the Holocene using the Greenland NGRIP ice core and selected auxiliary records. *J. Quat. Sci.*, *24*, 3-17.

Weyl, P. K. (1968). The role of the oceans in dramatic change: A theory of the ice ages, *Meteorol. Monogr.*, *8*, 37-62.

## Biographical Sketch

***Ken Munyikwa***

**Affiliation:** Centre for Science, Athabasca University

**Education:** MSc, Quaternary Engineering Geology; PhD, Quaternary Geology (Free University of Brussels, Belgium)

**Business Address:** Centre for Science, Athabasca University, 1 University Drive, Athabasca, AB T8N1T3, Canada

**Professional Appointments:** Associate Professor

In: Deglaciation
Editor: Marcella Boone

ISBN: 978-1-53612-501-6


*Chapter 2*

# THE INFLUENCE OF DEGLACIATIONS IN THE MID-LATITUDE EUROPEAN CLIMATE

***Gloria M. Martin-Garcia****

Department of Geology, University of Salamanca, Salamanca, Spain

## ABSTRACT

Global climate during the Quaternary has been deeply influenced by glacial-interglacial oscillations. Since the onset of glaciations, the Earth has experienced alternations between warm and stable climatic periods – coinciding with interglacials, and cold and highly variable climatic intervals – coinciding with glacials. In a suborbital timescale, climate oscillations were maximal during glacial onsets and, very especially, during deglaciations. It is very obvious that the extension of ice sheets has a direct influence on high-latitude climate, both in continents and oceans; contrary to this, climate in middle latitudes is not so influenced by global ice volume or the total extension of ice sheets, but by episodes of rapid variation of sea ice coverage. Since early Pleistocene the most pronounced stadial (cold) events in the mid-latitude North Atlantic occurred during deglaciations, both during glacial/interglacial transitions – Terminations – and episodes of lesser ice sheets depletion. These severe

* Corresponding author Email: gm.martin@usal.es.

cooling events were followed by very rapid, short-lived warming episodes that, when coinciding with Terminations of climatic cycles, initiated the climatic optimum of the following interglacial. Such millennial scale oscillations affected not only sea surface temperature, but also continental climate as far as the eastern Mediterranean basin. Most deglaciation onsets coincided with maxima in Northern Hemisphere summer insolation. Enhanced heat budget facilitated rapid ice sheets collapse and allowed the release of huge amounts of fresh water to the ocean, triggering the reduction or even the shutdown of North Atlantic deep water formation, which resulted in the advection of subpolar waters along the western European margin and the associated stadial. After deglaciation, once the freshwater perturbation stopped, the meridional overturning circulation rapidly resumed and warm surface water recirculated again in the North Atlantic, producing the fast temperature rise and initiation of interglacial climatic optimum across Europe.

**Keywords:** quaternary, deglaciations, climate change, continental climate, SST, North Atlantic, European climate

## ABBREVIATIONS

| | |
|---|---|
| AF | Arctic Front |
| AMOC | Atlantic meridional overturning circulation |
| AzC | Azores Current |
| AzF | Azores Front |
| AzH | Azores high-pressure area |
| HE | Heinrich events |
| NA | North Atlantic |
| NAC | North Atlantic Current |
| NADW | North Atlantic deep water |
| NAO | North Atlantic oscillation |
| $NAO^+$ | Positive phase of the NAO |
| $NAO^-$ | Negative phase of the NAO |
| ITCZ | Inter-tropical Convergence Zone |
| ENSO | El Niño/southern oscillation |

| | |
|---|---|
| SST | Sea surface temperature |
| T | Termination. |

# INTRODUCTION

## Climate Variation during the Quaternary

### *Glacial-Interglacial Oscillation*

During the Quaternary period global climate has been characterised by periodic oscillations of advance/retreat of the ice sheets that partly covered the Earth´s continents (e.g., Lisiecki and Raymo, 2007). Each climatic cycle begins with a very warm interval known as interglacial climatic optimum, when ice volume reaches a minimum value. Models suggest that some of these ice volume minima, like during marine isotope stage (MIS) 11 – from ~430 to ~365 ky, were even lower than present day values (Raynaud et al. 2003). The climatic optimum is followed by progressive cooling and increase of ice volume, marked by cool-warm oscillations of lesser amplitude and higher frequency than glacial-interglacial ones, which culminates with the highest extension of ice sheets during the cycle, known as glacial maximum. Each climatic cycle finishes with a rapid deglaciation, known as Termination, which connects with the following interglacial.

During the last million of years these climatic cycles have followed an approximate periodicity of 100 ky with superimposed climatic oscillations of ~41 ky and ~21 ky (e.g., Imbrie et al. 1992; Ruddiman, 2006; Huybers, 2007), all of which can be correlated with the orbital cycles described by Milankovitch (1941). Nevertheless, at a lower time scale, discrepancies exist between climatic records and orbital parameters, which suggest the influence of other factors in modulating the global climatic system. Some of these factors are thought to be the atmospheric concentration of greenhouse gases (Shackleton and Pisias, 1985; Kennett et al. 2000; Ruddiman, 2003, 2006; Toggweiler et al. 2006), oceanic circulation (Broecker et al. 1989; Raymo et al. 1997; McManus et al. 2004; Pisias et al. 2010), ice sheets size, and albedo effect (Imbrie et al. 1993; Denton el al., 2010).

### *Millennial-Timescale Climate Variation*

During middle and late Pleistocene several millennial-scale iceberg surges into the North Atlantic occurred that altered the ocean´s heat conveyor and caused fundamental reorganisations of the ocean-climate system, producing extreme cooling (Heinrich, 1988; Broecker, 1994, Cortijo et al. 1995; Hodell et al. 2008; Channel et al. 2012). These episodes, known as "Heinrich events" (Broecker et al. 1992), had a duration of a few hundred to a thousand years and a recurrence of ~7-10 ky (e.g., Bond and Lotti, 1995). Superimposed on this variability, a series of large shifts in air temperature, up to 15°C (Huber et al. 2006), in a few decades (Stuiver and Grootes, 2000), termed Dansgaard–Oeschger (D–O) oscillations have been registered in the Greenland ice-core records (e.g., Dansgaard et al. 1993; Grootes et al. 1993); the final event coincided with the Younger Dryas (e.g., Bond et al. 1992). These stadial-interstadial oscillations appear to be grouped in sequences of progressively cooler events, the end of each sequence marked by the deepest cooling; such sequences are known as "Bond cycles" (e.g., Broecker et al. 1990) and are thought to be linked to changes in the mode of overturning circulation in the Atlantic ocean (Broecker et al. 1985).

Most of Heinrich events happened at the end of the colder phase of a Bond cycle and were followed by warming to almost interglacial temperatures (e.g., Bond et al. 1992; Wright and Flower, 2002).

## Climate in the Subtropical and Middle Latitude North Atlantic and Western Europe

Although ocean dynamics (especially heat transport) are responsible for controlling sea surface temperature (SST) on interdecadal timescale, interannual variations are mainly driven by atmospheric heat flux changes (Bjerknes, 1964). In this sense, the climate regime over middle latitude

North Atlantic and Western Europe is mainly governed by the position and intensity of the Azores high-pressure area (AzH) and its latitudinal motion and intensification through the annual seasonal cycle. During spring/summer the AzH migrates northward, which results in strong northerly winds, while during fall/winter months the AzH is significantly weaker (Maze et al. 1997; Coelho et al. 2002). In consequence, this region is under the influence of the North Atlantic Oscillation (NAO) anomaly (Hurrell, 1995). This anomaly is measured by the NAO index, which records the difference in atmospheric pressure between the Iceland low and the AzH. The anomaly is positive ($NAO^+$) when the AzH is reinforced and the difference with the Iceland low is greater. In this scenario, westerly winds become stronger and the Gulf Stream migrates northward, which results in anomalously mild and wet winters for northern Europe, while southern Europe becomes anomalously cold and dry. The NAO also affects the sea ice coverage in the North Atlantic (Deser et al. 2000) and, indirectly, the thermohaline circulation (Dickson et al. 1996). In SW Europe, $NAO^+$ phases are associated with reinforcement of the seasonal upwelling off Portugal, via strengthening of NE trade winds. In $NAO^-$ conditions the westerlies are weaker, the Gulf Stream is located southward, climate in NW Europe is dry while in the south is wet and off Iberia the Azores Current prevails, inducing warm conditions and the reduction of the upwelling (Dickson et al. 1996).

In tropical regions, climatic conditions are mainly governed by the alternation of rainy (warm) and dry (colder) seasons that trace the seasonal motion of the ITCZ. The southward migration of the ITCZ appears to be linked to cooling on the North Atlantic over various time-scales (Black et al. 1999; Peterson et al. 2000; Schmidt et al. 2004) through weakening of the Atlantic Meridional Overturning Circulation (Schiller et al. 1997).

The climate of southern Europe, specifically the Iberian Peninsula, is also connected with the ENSO through more complex mechanisms. For example, both ENSO and NAO phenomena influence rainfall in Iberia (Rodo et al. 1997; Trigo et al. 2002). ENSO oscillation signals have been observed in the rainfall records of the past century in Iberia. There is clear evidence that the western part of the peninsula is more generally under the influence of NAO in winter while the eastern part is under the influence of ENSO in spring and autumn.

# INFLUENCE OF DEGLACIATIONS IN THE EUROPEAN ATLANTIC MARGIN CLIMATE

## Climate Variation in the Western European Margin during Deglaciations

The Western European margin is under the influence of the eastern gyre of the North Atlantic, that consists of two branches, the North Atlantic Current (NAC) in the north, and the Azores Current (AzC) in the south. The NAC constitutes a wide transition zone between the cold Arctic waters of high latitudes, and the warm subtropical waters of the AzC; the Artic Front (AF) and the Azores Front separate the three water masses. When the NAC reaches the continental margin, it diverts southwards and advects water of subpolar origin (~46°, McCartney and Talley, 1982; Branbilla and Talley, 2008) along mid-latitude and southern Europe – the Portugal Current –, until it meets the northward branch of the AzC – the Iberian Poleward Current.

During deglaciations, both Terminations and other important reductions of ice volume, profound reorganizations of North Atlantic circulation occurred that reduced the northward flow of warm water, and resulted in the most rapid and coldest events registered in the mid-latitude North Atlantic during the Quaternary, including the most severe glacial maxima. During some of these events the Southwestern Iberian margin registered drops of 8°C in the winter SST, that reached ~6.2°C (Martin-Garcia et al. 2015b) – today temperature being 15.5°C (Locarnini et al. 2010). The amplitude and duration of these cooling episodes were directly related with the magnitude of the ice volume reduction.

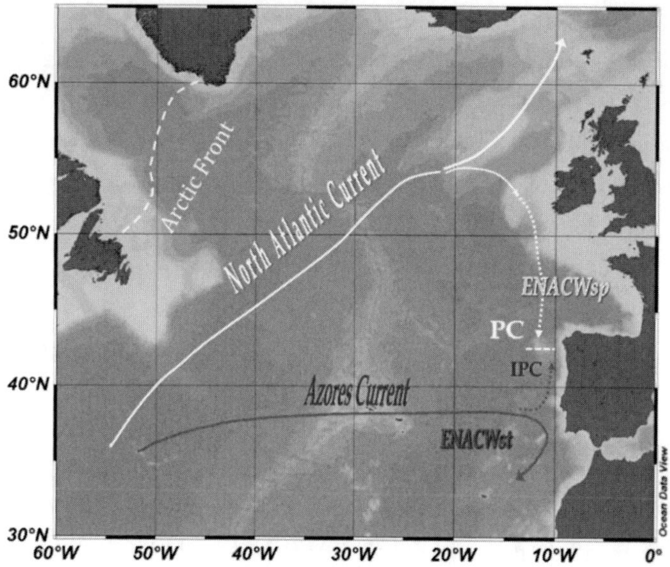


Figure 1. Map showing North Atlantic currents that bathe mid-latitude European margin. The surface limit between the PC and the IPC is according to Fiúza et al. (1998). The position of the Arctic Front is taken from Swift (1986) and Pflaumann et al. (2003). *PC* Portugal Current, *IPC* Iberian Poleward Current, *ENACWsp* Eastern North Atlantic Central Water of subpolar origin, *ENACWst* Eastern North Atlantic Central Water of subtropical origin.

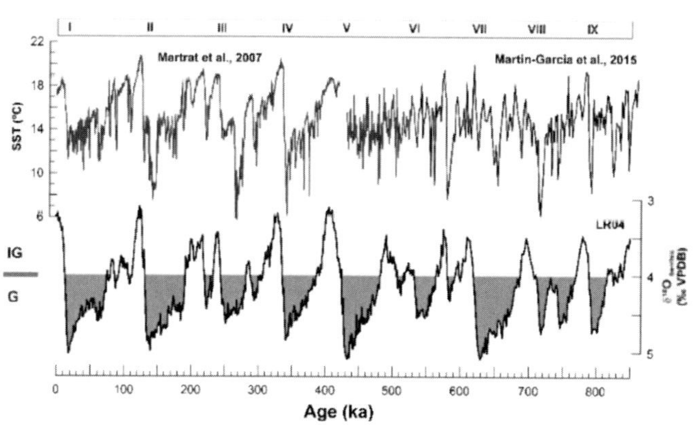


Figure 2. SST oscillation in the SW Iberian margin compared with global LR04 benthic stack of Lisiecki and Raymo (2005), generally used as ice volume indicator; filling enhances the ice volume threshold separating glacial (G) and interglacial (IG) stages. Annual SST from Martrat et al. (2007) is alkenone-based, and winter SST from Martin-Garcia et al. (2015c) is foraminifer-based. Roman numerals indicate Terminations.

Sudden and wide-amplitude warming – up to 11°C oscillation – followed the cooling and marked the climate optimum of the succeeding interglacial (Martin-Garcia et al. 2015b, c) (Figure 2).

To better understand the cause of these rapid climate oscillations, we have to consider different processes that occur during deglaciations and interact to influence surface and deep North Atlantic circulation and, in consequence, determine mid-latitude western European margin climate.

## Variation of the Arctic Front Position

Studies for the Quaternary have shown that surface water characteristics in the mid-latitude North Atlantic depend on the position of the oceanic fronts (e.g., Calvo et al. 2001; Naafs et al. 2010; Voelker et al. 2010). During glacial periods, the AF moved southward (e.g., Wright and Flower, 2002; Pflaumann et al. 2003; Martin-Garcia et al. 2015a) and even reached the northern Iberian margin during several peak events (Eynaud et al. 2009), allowing cold polar waters to expand into lower latitudes via recirculation through the subtropical gyre eastern currents. But such wide amplitude migrations of the AF especially occurred related to ice sheets instabilities in the northern Hemisphere (e.g., Bard et al. 2000) that produced freshwater fluxes to the high latitude ocean, which triggered perturbations in North Atlantic surface circulation and resulted in SST oscillations as far south as the Iberian margin (Lebreiro et al. 1996, 1997; Zahn et al. 1997; Cayre et al. 1999; de Abreu et al. 2005; Vautravers and Shackleton, 2006; Martrat et al. 2007; Eynaud et al. 2009; Rodrigues et al. 2011; Martin-Garcia et al. 2015b, c).

During glacial conditions, especially at times of glacial maxima, the southern position of the AF kept SST cold in most of North Atlantic and its continental margins. However, south of 37–39°N depending on the glacials, the European continental margin recorded fairly warm SST (Cayre et al. 1999; de Abreu et al. 2003; Vautravers and Shackleton, 2006;

Martrat et al. 2007; Rodrigues et al. 2011; Martin-Garcia et al. 2015b, c). This north-south gradient was caused by the eastwards diversion of the NAC produced by the advance of the AF, that blocked the northward flow of this current and, in consequence, the arrival of warm water to northern latitudes, while in the southernmost European margin the advection of warm surface water continued. The extremely southern position of the AF during some peak glacial conditions from the early Pleistocene to the Last Glacial Maximum, produced the maximal diversion of the NAC, which acquired an almost pure west-east direction (Pflaumann et al. 2003; Naafs et al. 2010; Martin-Garcia et al. 2015a). Only during deglaciations, major reorganizations of surface circulation occurred in the North Atlantic that reduced the northward flow of the NAC, promoting the southward migration of the eastern part of the AF and the expansion of cold subpolar waters along the southwestern European margin, trespassing the 37–39°N threshold of glacials.

After ice sheets collapse had finished and once the AF retreated northward again, the subtropical gyre resumed and the NAC invigorated, transporting warm surface water along the European margin, which resulted in the remarkable warming episode recorded after deglaciations. Prolonged and wide-amplitude deglaciations drastically reduced the extension of Northern Hemisphere ice sheets and allowed a northernmost position of the AF, which in turn, permitted the northward migration of the AzF and associated warm surface water that was recirculated by both the NAC and the AzC along the European margin. This sudden substitution of water masses bathing both high latitude and low latitude North Atlantic during and immediately after deglaciation, was registered by a very rapid change in the planktonic microfauna assemblage, which switched from polar species to warm (subtropical in south Portugal) assemblage (e.g., Ruddiman et al. 1989; Lebreiro et al. 1997; Wright and Flower, 2002; de Abreu et al. 2003, 2005; Martin-Garcia, 2015a, c).

## Weakening of the Atlantic Meridional Overturning Circulation (AMOC) and the European Margin climate

During deglaciations, North Atlantic circulation was severely affected not only in the surface, but also in depths. Both high and low latitudes registered slowdowns of the AMOC especially in Terminations, but also at other glacial-interglacial transitions (e.g., Wright and Flower, 2002; Hodell et al. 2008; Martin-Garcia et al. 2015c), which were caused by a weakened North Atlantic deep-water (NADW) formation and triggered by lower sea surface salinities in the subpolar North Atlantic as a consequence of fresh water releases at the onset of Northern Hemisphere ice sheets retreat. As the AMOC is responsible for ~50% of the total poleward advection of heat (Sabine et al. 2004; Adkins, 2013), during these events of reduced AMOC lower rates of heat were transferred to the North Atlantic and resulted in decreased SST from high to subtropical latitudes (e.g: Broecker et al. 1989; Stocker, 1999; McManus et al. 2004; Pisias et al. 2010; Martin-Garcia et al. 2015c).

The rate of AMOC is not the same in all deglaciations. During some of them, like Terminations I (Lippold et al. 2016) and II (Böhm et al. 2014), an active but shallow overturning cell maintained, exporting northern-sourced water at intermediate depth, the bottom of the North Atlantic being occupied by Antarctic Bottom Water, similar to the "cold mode" of AMOC that prevailed during glacial maxima (Boyle and Keigwin, 1987; Sarnthein et al. 1994). Strong sinking rate, although relatively superficial, can counteract the effect of deglaciation in decreasing SST, like during T VII, when an anomalously weak cooling was recorded in the southwestern Iberian margin (Martin-Garcia et al. 2015c) although this Termination was one of the largest deglaciations of the whole Quaternary.

Once the fresh water perturbation produced by ice sheets collapse finished, the NADW export abruptly resumed and the deep circulation switched to a northern-sourced regime (Böhm et al. 2014). The reactivation of the AMOC strengthened the surface circulation in the North Atlantic, allowing the recirculation of warm water along the European margin. Such rapid reorganization of oceanographic conditions resulted in the remarkable warming that followed deglaciations (e.g., Martin-Garcia, 2015a, c).

## Increased ice Rafting Events (Heinrich-Type Events)

Various studies have demonstrated the relationship between northern Hemisphere ice sheets instabilities that produced increased iceberg calving and dispersal, and cooling events in the southwestern European margin during the Quaternary. Such events were accompanied by high percentages of the polar species of foraminifer *Neogloboquadrina pachyderma* sinistral and layers of ice rafted debris (IRD). During most deglaciations, IRD layers were even recorded at the southernmost European margin (e.g., Lebreiro et al. 1996; Bard et al. 2000; Martrat et al. 2007; Rodrigues et al. 2011; Martin-Garcia et al. 2015c). Continuous massive surges and melting of iceberg during deglaciations increased the freshwater flow into the subpolar North Atlantic, which drastically reduced the export of NADW (e.g., Rühlemann et al. 1999; Wright and Flower, 2002; McManus et al. 2004), and resulted in the spread of polar waters and icebergs into mid-latitude North Atlantic and southwestern Europe, with the associated pronounced decrease of SST (Ruddiman et al. 1989; Naafs et al. 2011; Rodrigues et al. 2011; Martin-Garcia et al. 2015b, c).

Böhm et al. (2014) demonstrated that complete shutdowns of NADW formation, in the last glacial cycle, only occurred during Heinrich stadials close to glacial maxima with increased ice coverage. This mechanism could have worked also during the Pleistocene, as deduced from records of the southernmost Atlantic European margin, which registered most pronounced cooling associated to Heinrich-type events occurring during Terminations either with maximal ice coverage, like T IX, of very prolonged in time, like T VIII (Martin-Garcia et al. 2015c).

# INFLUENCE OF DEGLACIATIONS IN THE EUROPEAN INLAND CLIMATE

## Variation of NAO Index and Continental Climate during Deglaciations

The NAO is coupled to the trade winds responsible of a great part of inland climate, this index determining both the speed variability of the wind and winter precipitation levels (Hurrell, 1995; George and Saunders, 2001; Trigo et al. 2004). Observed trend towards a $NAO^+$ has been used to explain recent winter warming over the Northern Hemisphere continents (Hurrell 1996). Variations of the NAO index also have further consequences, triggered by changes in climate over continental Europe. In this line, the reduction in Mediterranean Sea level observed from the 1960s to the 1990s, has been explained, at least in a significant proportion, by the strengthening of the NAO over the period, which would produce atmospheric pressure anomalies combined with changes in evaporation and precipitation (Tsimplis and Josey, 2001).

To confidently extrapolate present-day ocean-ice-atmosphere interactions and interpret Quaternary climate dynamics, it is imperative to study lower than decadal frequency climate variability. In this line, several authors reconstructed historical variations of the NAO index and other atmospherical conditions, based on the study of tree-ring data (e.g., Cook et al. 1998; Glueck and Stockton, 2001), ice core data from Greenland (e.g., Barlow et al. 1993; White et al. 1996; Appenzeller et al. 1998), stalagmite growth rate (Proctor et al. 2000), or others (e.g., Cullen et al. 2000; Garcia et al. 2000; Luterbacher et al. 2001; Rodrigo et al. 2001). One of the most useful of these studies is that from Tremblay et al. (1997) using driftwood records and a dynamic-thermodynamic sea-ice model to reconstruct high latitude atmospheric circulation during the whole Holocene. Their results show that since the last deglaciation the average atmosphere-ocean system in the North Atlantic shifted abruptly between two main modes of centennial to millennial duration; the first one with negative NAO and large ice export, and the second one, more similar to present-day trend, with positive NAO, and low ice export. These swings in the atmosphere-ocean system state produced variations in the wind speed and direction, in the heat and humidity transfer between the North Atlantic and Europe and in the paths and intensity of storms; in summary, they were responsible of climatic variation in the Eurasian continent, recorded by proxies such as tree rings, historical records of precipitation, etc. When extrapolating to previous glacial/interglacial transitions, similar swifts in the climatological system situation must have happened, but the rapid changes were dampened by the slower response of vegetation communities, main proxy used to study past continental climate. Because of this, the response of continental climate to deglaciations is mainly recorded by a high-amplitude shift between the glacial-maximum-mode to the interglacial-optimum-mode (Tzedakis et al. 2006, 2009; Margari et al. 2010, 2014; Sánchez Goñi et al. 2016) although the transition between both situations probably occurred as a succession of centennial climatic oscillations, similar to the present interglacial, rather than as abrupt shifts.

## Influence of Sea Surface Temperature Variation in Continental Climate

Deglaciations resulted in very wide SST oscillations, as explained in previous sections. Such thermal variation did not only affect to the ocean surface, but also to inland climate via NAO variation. Numerous studies in present-day ocean (Barnston and Livezey, 1987; Peng and Mysak, 1993; Peng et al. 1995; Czaja and Frankignoul, 1999) clearly demonstrate that the NAO signal is connected with preceding SST variations in the North Atlantic, and even modulated by them at a timescale of ~60 years (Higuchi et al. 1999). This connection between North Atlantic SST anomalies and atmospheric circulation has been further supported by models (Rodwell et al. 1999; Mehta et al. 2000; Robertson et al. 2000), and explained (Rodwell et al. 1999) via evaporation, precipitation and atmospheric heating processes over the ocean that lead to variation of temperature and precipitation inland. In this line, variations of SST recorded in the North Atlantic during the last two centuries had important consequences in American and European climate, especially in precipitation rates (Enfield et al. 2001). Interannual changes in sea/land temperature co-variances are also related with NAO (Stephenson et al. 2000).

The Tropical and Subtropical North Atlantic is the most important region for this teleconnection ocean-inland climate, as its SST co-vary with the trade winds fluctuations (Nobre and Shukla, 1996; Black et al. 1999), influenced by shifts in the mean latitudinal position of the ITCZ, as fossil pollen records from southern Europe reveal (Tzedakis et al. 2009). Nevertheless, in a decadal timescale thermal variation in the subpolar ocean has also its consequence in inland climate, as the increase in SST in the Labrador Sea was observed to be paired with positive anomalies in air pressure over Greenland, that is, a weaker than normal Iceland Low and associated decrease of the NAO as a consequence of the southward diversion of the storm tracks (Wanner et al. 2001).

Whatever the driving mechanisms are, it has been observed that variations in percentage of the southwester European temperate vegetation closely mirrored the planktonic isotopic signal during the penultimate glaciation (Margari et al. 2014), which suggests the synchronous response of vegetation to North Atlantic SST variability. Temperate arboreal communities expanded coinciding with interstadial intervals recorded by both SST variations and Greenland ice cores, while steppe vegetation was dominant during cold intervals (stadials) (Sánchez Goñi et al. 2008; Margari et al. 2010). A similar covariance between SST and European continental climate signals was observed in speleothems from France (Wainer et al. 2013). During deglaciations, as consequence of large fresh water discharges, coldest SST were recorded in the SW European margin (Martin-Garcia et al. 2015b, c); such oscillations, via atmosphere-ocean coupling, influenced the continental climate in southern Europe, where moisture-requiring vegetation collapsed and was substituted by less demanding steppe communities (e.g., Roucoux et al. 2005; Combourieu et al. 2007; Sánchez Goñi et al. 2008, 2013; Margari et al. 2014).

## Influence of the Thermohaline Circulation Variation

Over the period 1968-1982, due to an increased continental freshwater input into the polar basin, a widespread freshening of subpolar waters occurred in the North Atlantic (the Great Salinity Anomaly) that deeply affected the NADW formation rate (Dickson et al. 1988; Mysak et al. 1990). Such freshwater fluxes in the area of deepwater formation have significant consequences in the global thermohaline circulation (Broecker et al. 1985; Aagard and Carmack, 1989; Stocker and Mysak, 1989) that, in turn, is responsible for controlling the heat transport from low to high latitudes (Boyle and Keigwin, 1987; Adkins, 2013), and modulates climate (Rahmstorf, 2002). Interdecadal variability of the NAO has been explained by variations in the thermohaline circulation (Timmermann et al. 1998). Rapid strengthening of the AMOC results in increasing of SST, which in turn, produces the strengthening of the NAO. One of the main characteristics of deglaciations is the very rapid resumption of NADW formation immediately after the melting-sourced fresh water input stops, which reactivates the advection of warm water of the subtropical gyre to high latitudes and results in rapid increase of SST over the North Atlantic (e.g., Martin-Garcia el al., 2015a, c). In this scenario, northern Europe inland climate would experience mild and wet winters during deglaciations, while southern Europe would become cold, dry and windy. This interpretation is supported by pollen records from the Balkan to the Iberian Peninsulas (Moreno et al. 2005; Combourieu et al. 2007) showing an increase of steppe vegetation at the expense of temperate forest during the last deglaciation, which indicates a climate shift towards colder and dryer conditions.

## Influence of Ice Sheets Variation and the Arctic Front Migration

Ice coverage influences the heat flux to the atmosphere and, in consequence, can modify the atmospheric circulation. Studies of trends in the modern ocean have shown that decadal sea ice variability is strongly coupled with the NAO index (Yi et al. 1999; Deser et al. 2000); particularly the area of sea ice coverage in Nordic and Baltic seas correlates negatively with the NAO index, while in the Labrador Sea it is positively correlated (Koslwski and Loewe, 1994; Koslowski and Glaser, 1999; Vinje, 2001). Although this NAO-sea ice coverage interaction is clear, it does not mean that European inland climate variability can be correlated to the present-day sea ice coverage. As a matter of fact, studies of southern European climate regimes during previous interglacials, measured as forest development, showed that forest extent was not related with residual ice volume but with other climate forcing, like the amplitude of insolation (Tzedakis, 2005). Nevertheless, during glacial intervals the correlation between ice sheets volume and forest extent was very close. This suggests that deglaciations influence inland climate by changing its main driving forcing and not merely by a switch between two different climate regimes.

The influence of ice volume on continental climate does not limit to the sea-ice extension and location. Global ice volume during glacial periods and especially the extension of Northern Hemisphere ice sheets, determine the strength of the following interglacial, that is, a very warm and rapid interglacial peak follows a strong (in ice volume and cold temperature) glacial maximum, and vice-versa; a greater level of deglaciation corresponding also to stronger glacials (Lang and Wolff, 2011). Such connection has been identified for all the deglaciations since 800 ka, not only for Terminations but also for other deglaciations occurring both during glacial and interglacial periods, and can be explained following the model of Parrenin and Paillard (2003) for deglaciations. According to these authors, deglaciations occur when astronomical forcing combined with ice sheet volume reach a threshold; during periods of weak forcing, a greater ice sheet extension is required for a deglaciation to occur. As glaciations initiate only as consequence of astronomical forcing, these periods of weak forcing will allow a greater deglaciation and, in consequence, a stronger interglacial.

Previous to deglaciations, inland climate in southern Europe was generally cold and dry, its climatic severity depending on the global ice volume and particularly on the North Hemisphere ice sheet extension (Tzedakis, 2005; Sánchez Goñi et al. 2016); relatively high percentages of semi-desert vegetation were recorded when ice volume was great, as during glacials MIS 12, MIS 6.4, MIS 4 and MIS 2 (Moreno et al. 2005; Fletcher et al. 2010; Margari et al. 2014), while less severe glacial periods were marked by higher tree percentage, which require warmer and wetter conditions. Once deglaciation had finished, climate in these regions became milder and more humid, the temperature rise depending mainly on the amplitude of the deglaciation, that is, inversely depending on the ice volume during the previous glacial period. Such changes occurred both during Terminations and other deglaciations, and were mainly recorded by the rapid increase of temperate and Mediterranean arboreal pollen percentage in Southern Europe (Tzedakis et al. 2006, 2009; Margari et al. 2010, 2014; Sánchez Goñi et al. 2016), but also by speleothems (Ayalon et al. 2002; Bard et al. 2002).

## Conclusion

During deglaciations, both Terminations and transitions from glacial to interglacial substages, profound reorganizations of North Atlantic circulation occurred that reduced the northward flow of warmth, promoting the southeastwards migration of the Arctic front and the expansion of cold subpolar waters along the European margin, trespassing the 37–39ºN threshold of glacials. This resulted in the deepest cooling registered in the region during the Quaternary, including the most severe glacial maxima. The amplitude and duration of these cooling episodes were directly related with the magnitude of the ice volume reduction. After ice sheets collapse had finished and once the AF retreated northward again, the subtropical gyre resumed and the NAC invigorated, advecting warm surface water along the European margin, which resulted in the Sudden and wide-amplitude warming – up to 11°C oscillation in the southwestern Iberian margin – recorded after deglaciations and that marked the climate optimum of the succeeding interglacial. This sudden substitution of water masses bathing both high and low latitude North Atlantic during and immediately after deglaciations, was registered by a very rapid change in planktonic microfauna assemblages, that switched from polar species to warm (subtropical in south Portugal) assemblage.

Decreased SST from high to subtropical latitude North Atlantic were the consequence of slowdowns of the AMOC caused by a weakened NADW formation and triggered by lower sea surface salinities in the subpolar North Atlantic as a consequence of fresh water releases at the onset of Northern Hemisphere ice sheets retreat. Nevertheless, the rate of AMOC was not the same in all deglaciations – during some of them, an active but shallow overturning cell maintained, exporting northern-sourced water at intermediate depth. This mechanism can counteract the effect of deglaciation in decreasing SST and be responsible for the anomalously weak cooling recorded during some of the largest deglaciations of the whole Quaternary, like T VII. Complete shutdowns of NADW formation only occurred during Heinrich-type events coinciding with deglaciations either with maximal ice coverage, like T IX, of very prolonged in time, like T VIII, when continuous massive surges and melting of iceberg increased the freshwater flow; such events resulted in the most pronounced cooling. Once the fresh water perturbation produced by ice sheets collapse finished, the NADW export abruptly resumed and the deep circulation switched to a northern-sourced regime. The reactivation of the AMOC strengthened the surface circulation in the North Atlantic. Such rapid reorganization of oceanographic conditions resulted in the remarkable warming that followed deglaciations.

The wide amplitude SST oscillation registered in the North Atlantic during deglaciations affected atmospheric circulation, heating processes over the ocean, evaporation and precipitation rates, which led to variation of temperature and precipitation inland. Covariance between SST and European continental climate signals was observed in speleothems from France and vegetation communities in southern Europe; during deglaciations continental climate in southern Europe was cold, dry and windy, which caused the substitution of temperate, moisture-requiring vegetation for less demanding steppe communities. Once deglaciation had finished, climate in these regions became milder and more humid, the temperature rise depending mainly on the amplitude of the deglaciation. Nevertheless, deglaciations influenced mid-latitude inland climate not merely by switching between two different climate regimes, but changing its main driving forcing. During glacials climatic severity in southern Europe depended on the global ice volume, the forest extent being closely correlated with the North Hemisphere ice sheet extension, while during interglacials forest coverage was not related with residual ice volume but with other climate forcing, like the amplitude of insolation.

## REFERENCES

Aagard, K. & Carmack, E. C. (1989). The role of sea ice and other fresh water in the arctic circulation.' *J. Geophys. Res.*, *94*, 14485–14498.

Adkins, J. (2013). The role of deep ocean circulation in setting glacial climates. *Paleoc.*, *28*, 539-561.

Appenzeller, C., Stocker, T. F. & Anklin, M. (1998). North Atlantic oscillation dynamics recorded in Greenland ice cores. *Sci.*, *282*, 446–449.

Ayalon, A., Bar-Matthews, M. & Kaufman, A. (2002). Climatic conditions during marine oxigen isotope stage 6 in the Eastern Mediterranean región from the isotopic composition of speleothems of Soreq Cave, Israel. *Geol.*, *30*, 303-306.

Bard, E., Antonioli, F. & Silenzi, S. (2002). Sea-level during the penultimate interglacial period based on a submerged stalagmite from Argentarola Cave (Italy). *Earth Planet. Sci. Lett.*, *196*, 135-146.

Bard, E., Rostek, F., Turon, J. L. & Gendreau, S. (2000). Hydrological impact of Heinrich Events in the Subtropical Northeast Atlantic. *Sci.*, *289*, 1321-1324.

Barlow, L. H., White, J. W. C., Barry, R. G., Rogers, J. C. & Grootes, P. M. (1993). The North At- lantic Oscillation signature in deuterium and deuterium excess signals in the Greenland ice sheet Project 2 ice core. *Geophys. Res. Lett.*, *20*, 2901–2904.

Barnston, A. G. & Livezey, R. E. (1987). Classification, seasonality and persistence of low frequency atmospheric circulation patterns. *Mon. Wea. Rev.*, *115*, 1083–1126.

Bjerknes, J. (1964). Atlantic air-sea interaction, *Adv. Geophys.*, 10, 1–82.

Black, D. E., et al. (1999). Eight centuries of North Atlantic Ocean variability. *Sci.*, *286*, 1709-1713.

Böhm, E., Lippold, J., Gutjahr, M., Frank, M., Blaser, P., Antz, B., Fohlmeister, J., Frank, N., Andersen, M. B. & Deininger, M. (2014). Strong and deep Atlantic meridional overturning circulation during the last glacial cycle. *Nat.*, *000*, 1-4.

Bond, G. C. & Lotti, R. (1995). Iceberg discharges into the North Atlantic on millennial time scales during the last glaciation. *Sci.*, *267*, 1005-1010.

Bond, G., Heinrich, H., Broecker, W., Labeyrie, L., McManus, J., Andrews, J., Huon, S., Jantschik, R., Clasen, S., Simet, C., Tedesco, K., Klas, M., Bonani, G. & Ivy, S. (1992). Evidence for massive discharges of icebergs into the North Atlantic Ocean during the last glacial period. *Nat.*, *360*, 245-249.

Boyle, E. & Keigwin, L. (1987). North Atlantic thermohaline circulation during the past 20.000 years linked to high-latitude surface temperature. *Nat.*, *330*, 35-40.

Brambilla, E. & Talley, L. D. (2008). Subpolar Mode Water in the northeastern Atlantic: 1. Averaged properties and mean circulation. *J. Geophys. Res.*, *113*, C04025.

Broecker, W. S. (1994). Massive iceberg discharges as triggers for global climate change. *Nat.*, *372*, 421-424.

Broecker, W. S., Peteet, D. M. & Rind, D. (1985). Does the ocean-atmosphere system have more than one stable mode of operation? *Nat.*, *315*, 21-26.

Broecker, W. S., Kennett, J. P., Flower, B. P., Teller, J. T., Trumbore, S., Bonani, G. & Wolfli, W. (1989). Routing of meltwater from the Laurentide Ice Sheet during the Younger Dryas cold episode. *Nat.*, *341*, 318-321.

Broecker, W. S., Bond, G., Millie, K., Geotges, B. & Willy, W. (1990). A salt oscillator in the glacial Atlantic?, 1, The concept. *Paleoceanography*, *5*, 469-477.

Broecker, W. S., Bond, G., Klas, M., Clark, E. & McManus, J. (1992). Origin of the nothern Atlantic´s Heinrich events. *Clim. Dyn.*, *6*, 265-273.

Calvo, E., Villanueva, J., Grimalt, J. O., Boelaert, A. & Labeyrie, L. (2001). New insights into the glacial latitudinal temperature gradients in the North Atlantic. Results from $U^{k'}_{37}$ sea surface temperatures and terrigenous inputs. *Earth Planet. Sci. Lett.*, *188* (3-4), 509-519.

Cayre, O., Lancelot, Y. & Vincent, E. (1999). Paleoceaanographic reconstructions from planktonic foraminifera ogg the Iberian Margin: Temperature, salinity and Heinrich events. *Paleocean.*, *14* (3), 384-396.

Channell, J. E. T., Hodell, D. A., Romero, O., Hillaire-Marcel, C., de Vernal, A., Stoner, J. S., Mazaud, A. & Roehl. (2012). IODP Site U1302-U1303 (Orphan Knoll): Correlation of Brunhes detrital-layer stratigraphy into the North Atlantic. *Earth Planet. Sci. Letts.*, 317-318, 218-230.

Coelho, H., Neves, T., White, M., Leitpo, P. & Santos, A. (2002). A model for ocean circulation on the Iberian coast. *J. Mar. Sys.*, *32*, 181-198.

Combourieu Nebout, N., Bordon, A., Peyron, O., Kageyama, M. & Cazet, J. P. (2007). Mediterranean climate during the short-time events of the last Deglaciation and the Holocene: seasonality and gradient according to vegetation changes. *Geoph. Res. Abs.*, *9*, 07575.

Cook, E. R., D'Arrigo, R. D. & Briffa, K. R. (1998). The North Atlantic Oscillation and its expression in circum-Atlantic tree-ring chronologies from North America and Europe. *Holoc.*, *8*, 9–17.

Cortijo, E., Yiou, P., Labeyrie, L. & Cremer, M. (1995). Sedimentary record of rapid climatic variability in the North Atlantic Ocean during the last glacial cycle. *Pleocean.*, *10*, 911-926.

Cullen, H. M., D'Arrigo, R., Cook, E. R. & Mann, M. E. (2000). Multiproxy reconstructions of the North Atlantic Oscillation. *Paleocean*, *16*, 27–39.

Czaja, A. & Frankignoul, C. (1999). Influence of the North Atlantic SST on the atmospheric circulation, *Geophys. Res. Lett.*, *26*, 2969–2972.

Dansgaard, W., Johsen, S. J., Clauser, H. B., Kahl, J., Gundestrup, N. S., Hommer, C. U., Huidberg, C. S., Steffensen, J. P., Svernbjornsdottir, A. E., Jouzel, J. & Bond, G. (1993). Evidence for general instability of past climate from a 250 kyr ice-core record. *Nat.*, *364*, 218-220.

de Abreu, L., Shackleton, N. J., Schönfeld, J., Hall, M. A. & Chapman, M. (2003). Millennial-scale oceanic climate variability off the western Iberian margin during the last two glacial periods. *Mar. Geol.*, *196*, 1-20.

de Abreu, L., Abrantes, F., Shackleton, N. J., Tzedakis, P. C., McManus, J. F., Oppo, D. W. & Hall, M. A. (2005). Ocean climate variability in the eastern North Atlantic during interglacial marine isotope stage 11: A partial analogue to the Holocene? *Paleocean.*, *20*.

Denton, G. H., Anderson, R. F., Toggweiler, J. R., Edwards, R. L., Schaefer, J. M. & Putnam, A. E. (2010). The Last Glacial Termination. *Sci.*, *328*, 1652-1656.

Deser, C., Walsh, J. El. & Timlin, M. S. (2000). Arctic sea ice variability on the context of recent atmospheric circulation trends. *J. Clim.*, *13*, 617-633.

Dickson, R. R., Meincke, J., Malmberg, S. A. & Lee, A. J. (1988). The Great Salinity Anomaly in the northern North Atlantic 1968–1982. *Prog. Ocean.*, *20*, 103–151.

Dickson, R., Lazier, J., Meinke, J., Thines, P. & Swift, J. (1996). Long term coordinated changes in the convective activity of the North Atlantic. *Progr. Oceanogr.*, *38*, 241-295

Enfield, D. B., Mestas-Nunez, A. M. & Trimble, P. J. (2001). The Atlantic Multidecadal Oscillation and its relationship to rainfall and river flows in the Continental U.S.. *Geophys. Res. Lett.*, *28*, 2077–2080.

Eynaud, F., de Abreu, L., Voelker, A., Schönfeld, J., Salgueiro, E., Turon, J. L., Penaud, A., Toucanne, S., Naughton, F., Sanchez-Goñi, M. F., Malaizé, B. & Cacho, I. (2009). Position of the Polar Front along the western Iberian Margin during key cold episodes of the last 45 ka. *Geochem. Geophys. Geosyst.*, *10*, Q07U05.

Fiúza, A. F. G., Hamann, M., Ambar, I., Díaz del Río, G., González, N. & Cabanas, J. M. (1998). Water masses and their circulation off western Iberia during May 1993. *Deep-Sea Res.*, *45*, 1127-1160.

Fletcher, W. J., Sánchez Goñi, M. F., Allen, J. R. M., Cheddadi, R., Combourieu-Nebout, N., Huntley, B., Lawson, I., Londeix, L., Magri, D., Margari, V., Müller, U. C., Naughton, F., Novenko, E., Roucoux, K. & Tzedakis, P. C. (2010). Millennial-scale variability during the last glacial in vegetation records from Europe. *Quat. Sci. Rev.*, *29*, 2839-2864.

Garcia, R., Gimeno, L., Hernandez, E., Prieto, R. & Ribera, P. (2000). Reconstructing the North Atlantic atmospheric circulation in the 16th, 17th and 18th centuries from historical sources. *Clim. Res.*, *14*, 147–151.

George, S. E. & Saunders, M. A. (2001). North Atlantic Oscillation impact on tropical north Atlantic winter atmospheric variability, *Geophy. Res. Lett.*, *28*, 1015–1018.

Glueck, M. F. & Stockton, C. W. (2001). Reconstruction of the North Atlantic Oscillation. *Int. J. Climatol.*, *21*, 1453–1465.

Grootes, P. M., Stuiver, M., White, J. W. C., Johnsen, S. & Jouzel, J. (1993). Comparison of oxigen isotope records from the GISP2 and GRIP Greenland ice cores. *Nat.*, *255*, 522-554.

Heinrich, H. (1988). Origin and consequences of cyclic ice rafting in the Northeast Atlantic Ocean during the past 130,000 yrs. *Quat. Res.*, *29*, 142-152.

Higuchi, K., Huang, J. & Shabbar, A. (1999). A wavelet characterization of the North Atlantic oscillation variation and its relationship to the North Atlantic sea surface temperature. *Int. J. Climatol.*, *19*, 1119–1129.

Hodell, D. A., Channell, J. E. T., Curtis, J. H., Romero, O. E. & Röhl, U. (2008). Onset of "Hudson Strait" Heinrich events in the eastern North Atlantic at the end of the middle Pleistocene transition (~640 ka)? *Paleocean.*, *23*, doi:10.1029/2008PA001591.

Huber, C., Leuenberger, M., Spahni, R., Flückiger, J., Schwander, J., Stocker, T. F., Johnsen, S., Landais, A. & Jouzel, J. (2006). Isotope calibrated Greenland temperature record over marine isotope stage 3 and its relation to $CH_4$. *Earth Plan. Sci. Lett.*, *243* (3-4), 504-519.

Hurrell, J. W. (1995). Decadal trends in the North Atlantic Oscillation: Regional temperaturas and precipitation. *Sci.*, *269*, 676-679.

Hurrell, J. W. (1996). Influence of variations in extratropical wintertime teleconnections on northern hemisphere temperature. *Geoph. Res. Lett.*, *23*, 665-668.

Huybers, P. (2007). Glacial variability over the last two million years: an extended depth-derived agemodel, continuous obliquity pacing and the Pleistocene progression. *Quat. Sci. Rev.*, *26*, 37-55.

Imbrie, J., Boyle, E. A., Clemens, S. C., Kutzbach, J., Martinson, D. G., McIntyre, A., Mix, A. C., Molfino, B., Morley, J. J., Peterson, L. C., Pisias, N. G., Prell, W. L., Raymo, M. E., Shackleton, N. J. & Toggweiler, J. R. (1992). On the Structure and Origin of Major Glaciation Cycles 1. Linear responses to Milankovitch forcing. *Paleocean.*, *7*, 701-738.

Imbrie, J., Berger, A., Boyle, E. A., Clemens, S. C., Duffy, A., Howard, W. R., Kukla, G., Kutzbach, J., Martinson, D. G., McIntyre, A., Mix, A. C., Molfino, B., Morley, J. J. Peterson, L. C, Pisias, N. G., Prell, W. L., Raymo, M. E., Shackleton N. J. & Toggweiler, J. R. (1993). On the structure and origin of major glaciation cycles: 2. The 100,000-year cycle. *Paleocean.*, *8*, 699-735.

Kennett, J. P., Cannariato, K. G., Hendy, I. L. & Behl, R. J. (2000). Carbon isotopic evidence for methane hydrate instablity during Quaternary interstadials. *Sci.*, *288* (5463), 128-133.

Koslowski, G. & Loewe, P. (1994). The western Baltic sea ice season in terms of a mass-related severity index: 1879–1992, Part I: Temporal variability and association with the North Atlantic Oscillation. *Tellus*, *46A*, 66–74.

Koslowski, G. & Glaser, R. (1999). Variations in reconstructed ice winter severity in the Western Baltic from 1501 to 1995, and their implications for the North Atlantic Oscillation. *Clim. Chan.*, *41*, 175–191.

Lang, N. & Wolff, E. W. (2011). Interglacial and glacial variability from the last 800 ka in marine, ice and terrestrial archives. *Clim. Past*, *7*, 361-380.

Lebreiro, S. M., Moreno, J. C., McCave, I. N. & Weaver, P. P. E. (1996). Evidence for Heinrich layers off Portugal (Tore Seamount: 39ºN, 12ºW). *Mar. Geol.*, *131*, 47-56.

Lebreiro, S. M., Moreno, F. C., Abrantes, F. & Pflaumann, U. (1997). Productivity and paleoceanographic implications on the Tore Seamount (Iberian Margin) during the last 225 kyr: Foraminiferal evidence. *Paleocean.*, *12*, 5, 718-727.

Lippold, J., Gutjahr, M., Blaser, P., Christner, E., de Carvalho Ferreira, M. L., Mulitza, S., Chistl, M., Wombacher, F., Böhm, E., Antz, B., Cartapanis, O., Vogel, H. & Jaccard, S. L. (2016). Deep water provenance and dynamics of the (de)glacial Atlantic meridional overturning circulation. *Earth Planet. y Sci.e lett.*, *445*, 68-78.

Lisiecki, L. E. & Raymo, M. E. (2005). A Pliocene-Pleistocene stack of 57 globally distributed benthic d$^{18}$O records. *Paleocean.*, *20*, PA1003.

Lisiecki, L. E. & Raymo, M. E. (2007). Plio-Pleostocene climate evolution: trends and transitions in glacial cycle dynamics. *Quat. Sci. Rev.*, *26*, 56-69.

Locarnini, R. A., Mishonov, A. V., Antonov, J. I., Boyer, T. P., Garcia, H. E., Baranova, O. K., Zweng, M. M. & Johnson, D. R. (2010). World Ocean Atlas 2009, Volume 1: Temperature. S. Levitus, Ed. NOAA Atlas NESDIS 68, U.S. *Govern. Print. Off.*, Washington, D.C., 184 pp.

Luterbacher, J., Xoplaki, E., Dietrich, D., Rickli, R., Jacobeit, J., Beck, C., Gyalistras, D., Schmutz, C. & Wanner, H. (2001). Reconstruction of sea level pressure fields over the eastern North Atlantic and Europe back to 1500, *Clim. Dyn.*

Margari, V., Skinner, L. C., Tzedakis, P. C., Ganopolski, A., Vautravers, M. & Shackleton, N. J. (2010). The nature of millennial-scale climate variability during the past two glacial periods. *Nat. Geosc.*, *3*, 127–131.

Margari, V., Skinner, L. C., Hodell, D. A., Martrat, B., Toucanne, S., Grimalt, J. O., Gibbard, P. L., Lunkka, J. P. & Tzedakis, P. C. (2014). Land-ocean changes on orbital and millennial time scales and the penultimate glaciation. *Geol.*, *42*, 183-186.

Martin-Garcia, G. M., Sierro, F. J. & Flores, J. A. (2015a). Reconstruction of North Atlantic circulation affecting the SW Iberian Margin during glacial stages from 840 to 530 ka. *Proc. VIII Con. MIA*, *15*, 237-240.

Martin-Garcia, G. M., Sierro, F. J., Hodell, D. A. & Flores, J. A. (2015b). Climate oscillations on the SW Iberian Margin from 860 to 490 ka. *Proc. VIII Con. MIA*, *15*, 429-432.

Martin-Garcia, G. M., Alonso-Garcia, M., Sierro, F. J., Hodell, D. A. & Flores, J. A. (2015c). Severe cooling episodes at the onset of deglaciations on the Southwestern Iberian margin from MIS 21 to 13 (IODP site U1385). *Glob. Planet. Chan.*, *135*, 159-169.

Martrat, B., Grimalt, J. O., Shackleton, N. J., de Abreu, L., Hutterli, M. A. & Stocker, T. F. (2007). Four Climate Cycles of Recurring Deep and Surface Water Destabilizations on the Iberian Margin. *Sci.*, *317* (5837), 502-507.

Maze, J. P., Arhan, M. & Mercier, H. (1997). Volume budget of the eastern boundary layer off the Iberian Peninsula. Deep Sea Research Part I: *Oceanog. Res. Pap.*, *44*, 1543-1574.

McCartney, M. S. & Talley, L. D. (1982). The Subpolar Mode Water of the North Atlantic Ocean. *J. Phys. Oceanogr.*, *12*, 1169-1188.

McManus, J. F., Francois, R., Gherardi, J., Keigwin, L. & Brown-Leger, S. (2004). Collapse and rapid resumption of Atlantic meridional circulation linked to deglacial climate change. *Nat.*, *428*, 834-837.

Mehta, V. M., Suarez, M. J., Manganello, J. V. & Delworth, T. L. (2000). Oceanic influence on the North Atlantic Oscillation and associated Northern Hemisphere climate variations: 1959–1993, *Geophy. Res. Lett.*, *27*, 121–124.

Milankovitch, M. (1941). Canon of insolation and the ice-age problem. *Roy. Serb. Sci. Special Publ. Sect. Math. Nat. Sci.*, *33*, 633.

Moreno, A., Cacho, I., Canals, M., Grimalt, J. O., Sánchez-Goñi, M. F., Shackleton, N. & Sierro, F. J. (2005). Links between marine and atmospheric processes oscillating on a millennial time-scale. A multi-proxy study of the last 50,000 yrs from the Alboran Sea (Western Mediterranean Sea). *Quat. Sci. Rev.*, *24*, 1623–1636.

Mysak, L. A., Manak, D. K. & Marsden, R. F. (1990). Sea-ice anomalies observed in the Greenland and Labrador Seas during 1901–1984 and their relation to an interdecadal Arctic climate cycle. *Clim. Dyn.*, *5*, 111–133.

Naafs, B. D. A., Stein, R., Hefter, J., Khélifi, N., De Schepper, S. & Haug, G. H. (2010). Late Pliocene changes in the North Atlantic Current. *Earth Planet. Sci. Lett.*, *298*, 434-442.

Naafs, B. D. A., Hefter, J., Ferretti, P., Stein, R. & Haug, G. H. (2011). Biomarker abundance in MIS 16 of IODP Site 306-U1313. doi:10.1594/PANGAEA.757948.

Nobre, P. & Shukla, J. (1996). Variations of sea surface temperature, wind stress, and rainfall over the tropical Atlantic and South America. *J. Clim.*, *9*, 2464–2479.

Parrenin, F. & Paillard, D. (2003). Amplitude and phase of glacial cycles from a conceptual model. *Earth Planet. Sc. Lett.*, *214*, 243–250.

Peng, S. & Mysak, L. A. (1993). A teleconnection study of interannual sea surface temperature fluctuations in the North Atlantic and precipitation and runoff over western Siberia. *J. Clim.*, *6*, 876–885.

Peng, S., Mysak, L. A., Ritchie, H., Derome, J. & Dugas, B. (1995). The differences between early and midwinter atmospheric responses to sea surface temperature anomalies in the northwest Atlantic. *J. Clim.*, *8*, 137–157.

Peterson, L. C., Haug, G. H., Hughen, K. A. & Tohl, U. (2000). Rapid changes in the tropical hydrologic cycle of the tropical Atlantic during the last glacial. *Sci.*, *290*, 1947-1951.

Pflaumann, U., Sarnthein, M., Chapman, M., de Abreu, L., Funnell, B., Huels, M., Kiefer, T. & Maslin, M. (2003). Glacial North Atlantic: sea-surface conditions reconstructed by GLAMAP 2000. *Paleocean.*, *18*, 1065.

Pisias, N. G., Clark, P. U. & Brook, E. J. (2010). Modes of global climate variability during Marine Isotope Stage 3 (60-26 ka). *J. Clim.*, *23* (6), 1581-1588.

Proctor, C. J., Baker, J., Barnes, W. L. & Gilmore, M. A. (2000). A thousand year speleothem proxy record of North Atlantic climate from Scotland. *Clim. Dyn.*, *16*, 815–820.

Rahmstorf, S. (2002). Ocean circulation and climate during the past 120,000 years. *Nat.*, *419*, 207-214.

Raymo, M. E., Oppo, D. W. & Curry, W. V. (1997). The mid-Pleistocene climate transition: A deep sea carbon isotopic perspective. *Paleocean.*, *12*, 546-559.

Raynaud, D., Loutre, M. F., Ritz, C., Chappellaz, J., Barnola, J. M., Jouzel, J., Lipenkov, V. Y., Petit, J. R. & Vimeux, F. (2003). Marine Isotope Stage (MIS) 11 in the Vostok ice core: CO2 forcing and stability of East Antarctica, In: Droxler, A. W., Poore, R. Z., Burkle, L. H. (Eds.) Earth´s Climate and Orbital Eccentricity. The marine Isotope Stage 11 Question. *Geophys. Monog.*, *137*. American Geophysical Union, Washington D. C., pp. 27-40.

Robertson, A. W., Mechoso, C. R. & Young-Joon, K. (2000). The influence of Atlantic sea surface temperature anomalies on the North Atlantic Oscillation, *J. Clim.*, *13*, 122–138.

Rodo, W., Baert, E. & Comin, F. A. (1997). Variations in seasonal rainfall in southern Europe during the present century: Relationships with the North Atlantic Oscillation and the El Niño-Southern Oscillation. *Clim. Dyn.*, *13*, 275-284.

Rodrigues, T., Voelker, A. H. L., Grimalt, J. O., Abrantes, F. & Naughton, F. (2011). Iberian Margin sea surface temperature during MIS15 to 9 (580-300 ka): Glacial suborbital variability versus interglacial stability. *Paleocean.*, *26*, 1-16.

Rodrigo, F. S., Pozo-Vázquez, D., Esteban-Parra, M. J. & Castro-Diez, Y. (2001). A reconstruction of the winter North Atlantic Oscillation Index back to AD 1501 using documentary data in southern Spain. *J. Geophys. Res.*, *106*, 14805–14818.

Rodwell, M. J., Rowell, D. P. & Folland, C. K. (1999). Oceanic forcing of the wintertime North Atlantic Oscillation and European climate, *Nat.*, *398*, 320–323.

Roucoux, K. H., de Abreu, L., Shackleton, N. J. & Tzedakis, P. C. (2005). The response of NW Iberian vegetation to North Atlantic climate oscillations during the last 65 kyr. *Quat. Sci. Rev.*, *24*, 1637-1653.

Ruddiman, W. F. (2003). Orbital insolation, ice volume and greenhouse gases. *Quat. Sci. Rev.*, *22* (15-17), 1597-1629.

Ruddiman, W. F. (2006). Ice-driven $CO_2$ feedback on ice volume. *Clim. Past*, *2*, 43-55.

Ruddiman, W. F., Raymo, M. E., Martinson, D. G., Clement, B. M. & Backman, J. (1989). Sea surface temperature reconstruction of DSDP Site 94-607 in the North Atlantic. doi:10.1594/PANGAEA.52373.

Rühlemann, C., Mulitza, S., Müller, P. J., Wefer, G. & Zahn, R. (1999). Warming of the tropical Atlantic Ocean and slowdown of thermohaline circulation during the last deglaciation. *Nat.*, *402*, 511-514.

Sabine, C. L., Feely, R. A., Gruber, N., Key, R. M., Lee, K., Bullister, J. L., Wanninkhof, R., Wong, C. S., Wallace, D. W. R., Tilbrook, B., Millero, F. J., Peng, T. H., Kozyr, A., Ono, T. & Rios, A. F. (2004). The oceanic sink for anthropogenic $CO_2$. *Sci.*, *305*, 367-371.

Sánchez Goñi, M. F., Landais, A., Fletcher, W. J., Naughton, F., Desprat, S. & Duprat, J. (2008). Contrasting impacts of Dansgaard-Oeschger events over a western European latitudinal transect modulated by orbital parameters. *Quat. Sci. Rev.*, *27*, 1136-1151.

Sánchez Goñi, M. F., Bard, E., Landais, A., Rossignol, L. & d'Errico, F. (2013). Air-sea temperature decoupling in western Europe during the last interglacial-glacial transition. *Nat. Geosci.*, doi:10.1038/ NGEO1924.

Sánchez Goñi, M. F., Llave, E., Oliveira, D., Naughton, F., Desprat, S., Ducassou, E., Hodell, D. A. & Hernández-Molina, F. J. (2016). Climate changes in south western Iberia and Mediterranean Outflow variations during two contrasting cycles of the last 1 Myrs: MIS 31–MIS 30 and MIS 12–MIS 11. *Glob. Planet. Chan.*, *136*, 18-29.

Sarnthein, M., Winn, K., Jung, S. J. A., Duplessy, J. C., Labeyrie, L., Erlenkeuser, H. & Ganssen, G. (1994). Changes in east Atlantic deepwater circulation over the last 30,000 years: eight time slice reconstructions. *Paleocean.*, *9*, 209-267.

Schieller, A., Milkolajewicz, U. & Voss, R. (1997). The stability of North Atlantic thermohaline circulation in a coupled ocean-atmosphere general circulation model. *Clim. Dyn.*, *13*, 325-347.

Schmidt, M. W., Spero, H. J. & Lea, D. W. (2004). Links between salinity variation in the Caribbean and North Atlantic thermohaline circulation. *Nat.*, *428*, 160-163.

Shackleton, N. J. & Pisias, N. G. (1985). Atmospheric carbon dioxide, orbital forcing and climate. In: *The Carbon cycle and atmospheric $CO_2$: natural variations, Archean to present*; Sundquist, E.T. and Broecker, W.S.; Ed; Am. Geoph. Un., Washington, D.C.

Stephenson, D. B., Pavan, V. & Bojariu, R. (2000). Is the North Atlantic Oscillation a random walk? *Int. J. Climatol.*, *20*, 1–18.

Stocker, T. F. (1999). Past and future reorganisations in the climate system. *Quat. Sci. Rev.*, *19*, 301-319.

Stocker, T. F. & Mysak, L. A. (1989). Climatic fluctuations on the century timescale; a review of high-resolution proxy-data. *Clim. Res. Gr. Rep.*, 89-13, Dept. Meteorol., McGill University, Montreal.

Stuiver, M. & Grootes, P. M. (2000). GISP2 oxygen isotope ratios. *Quat. Res.*, *53* (3), 277-284.

Timmermann, A. M., Latif, M., Voss, R. & Grótzner, A. (1998). Northern hemispheric interdecadal variability: a coupled air-sea mode, *J. Clim.*, *11*, 1906–1931.

Toggweiler, J. R., Russell, J. L. & Carson, S. R. (2006). Midlatitude westerlies, atmospheric CO2 and climate change during the ice ages. *Paleocean.*, *21* (2).

Tremblay, L. B., Mysak, L. A. & Dyke, A. S. (1997). Evidence from driftwood records for century- to-millennial scale variations of the high latitude atmospheric circulation during the Holocene. *Geophys. Res. Lett.*, *24*, 2027–2030.

Trigo, R. M., Osborn, T. J. & Corte-Real, J. M. (2002). The North Atlantic Oscillation influence on Europe: climate impacts and associated physical mechanisms. *Clim. Res.*, *20*, 9-17.

Trigo, R. M., Pozo-Vázquez, D., Osborn, T. J., Castro-Díez, Y., Gámiz-Fortis, S. & Esteban-Parra, M. J. (2004). North Atlantic oscillation influence on precipitation, river flow and water resources in the Iberian Peninsula. *Int. J. Climatol.*, *24*, 925–944.

Tsimplis, M. N. & Josey, S. A. (2001). Forcing of the Mediterranean Sea by atmospheric oscillations over the North Atlantic, *Geophys. Res. Lett.*, *28*, 803–806.

Tzedakis, P. C. (2005). Towards an understanding of the response of southern European vegetation to orbital and suborbital climate variability. *Quat. Sci. Rev.*, *24*, 1585–1599.

Tzedakis, P. C., Hooghiemstra, H. & Palike, H. (2006). The last 1.35 million years at Tenaghi Philippon: revised chronostratigraphy and long-term vegetation trends. *Quat. Sci. Rev.*, *25*, 3416-3430.

Tzedakis, P. C., Pälike, H., Roucoux, K. H. & de Abreu, L. (2009). Atmospheric methane, southern European vegetation and low-mid latitude links on orbital and millennial timescales. *Earth Planet. Sci. Lett.*, *277*, 307-317.

Vinje, T. (2001). Anomalies and trends of sea-ice extent and atmospheric circulation in the Nordic Seas during the period 1864-1998. *J. Clim.*, *14*, 255-267.

Vautravers, M. J. & Shackleton, N. J. (2006). Centennial-scale surface hydrology off Portugal during marine isotope stage 3: Insights from planktonic foraminiferal fauna variability. *Paleocean.*, *21*, PA3004.

Voelker, A. H. L., Rodrigues, T., Billups, K., Oppo, D., McManus, J., Stein, R., Hefter, J. & Grimalt, J. O. (2010). Variations in mid-latitude North Atlantic surface water properties during the mid-Brunhes (MIS 9-14) and their implications for the thermohaline circulation. *Clim. Past*, *6*, 531-552.

Wainer, K., Genty, D., Blamart, D., Bar-Matthews, M., Quinif, Y. & Plagnes, V. (2013). Millennial climatic instability during penultimate glacial period recorded on a south-western France speleothem. *Palaeog, Palaeocl. Palaeoec.*, *376*, p. 122–131.

Wanner, H., Brönnimann, S., Casty, C., Gyalistras, D., Luterbacher, J., Schmutz, C., Stephenson, D. B. & Xoplaki, E. (2001). North Atlantic oscillation-concepts and studies. *Surv. Geoph.*, *22*, 321-382

White, J. W. C., Gorodetzky, D., Cook, E. R. & Barlow, L. K. (1996). Frequency analysis of an annu- ally resolved, 700 year paleoclimate record from the GISP2 ice core; In: *Climate Variations and Forcing Mechanisms of the Last 2000 Years*; R. S. Bradley et al.; Ed; Springer-Verlag, New York, pp. 193–213.

Wright, A. K. & Flower, B. P. (2002). Surface and deep ocean circulation in the subpolar North Atlantic during the mid-Pleistocene revolution. *Paleocean.*, *17* (4), 1068.

Yi, D., Mysak, L. A. & Venegas, S. A. (1999). Decadal-to-interdecadal fluctuations of Arctic sea-ice cover and the atmospheric circulation during 1954–1994. *Atmos. – Oc.*, *37*, 389–415.

Zahn, R., Schönfeld, J., Kudrass, H. R., Park, M. H., Erlenkeuser, H. & Grootes, P. (1997). Thermohaline instability in the North Atlantic during meltwater events: Stable isotope and ice-rafted detritus records from core SO75-26KL, Portuguese margin. *Paleoc.*, *12*, 696-710.

## BIOGRAPHICAL SKETCH

***Gloria M. Martin-Garcia***

**Affiliation:** Department of Geology, University of Salamanca. Salamanca, Spain.

**Education:** Degree and PhD in Geology. Mastership in Geophysics. Degree and Mastership in Environmental Science. Mastership in Education.

**Business Address:** Dpto. de Geología (Área de Paleontología). Plaza de los Caídos s/n. 37008-Salamanca (Spain)

**Research Experience:** Study of foraminifers, Quaternary climate and oceanography.

**Professional Experience:** Environmental engineering and Education.

In: Deglaciation
Editor: Marcella Boone

ISBN: 978-1-53612-501-6


*Chapter 3*

# DEGLACIATION OF THE LATE WEICHSELIAN BARENTS SEA ICE SHEET

***Ivar O. Murdmaa and Elena V. Ivanova******
P. P. Shirshov Institute of Oceanology,
Russian Academy of Sciences, Moscow, Russia

## ABSTRACT

According to the prevailing view, the Barents Sea shelf and surrounding archipelagoes were covered by a grounded ice sheet united with the Scandinavian ice sheet during the Last Glacial Maximum (LGM). Its decay started in the deep marginal troughs, the Bear Island Trough in the southwest, Kvitøya and Franz Victoria troughs in the north, ca. 19-18 cal ka BP with development of ice streams and related iceberg calving over the shelf break. Sediment cores and shallow boreholes recovered two distinct units of glaciomarine sediments. The lower Unit III is represented by a glaciomarine diamicton, a disordered mixture of coarse-grained iceberg rafted debris (from sand to pebbles and boulders), and very fine-grained clay (dominated by <0.001 mm fraction) released from the melting glacier ice. It overlies directly the eroded Mesozoic bedrock surface or non-marine basal moraine deposits (glacial till) in

* Corresponding Author Email: e_v_ivanova@ocean.ru.

shelf depressions. The diamicton was deposited during the early deglaciation stage, which was apparently diachronous in the study area. The faunal and stable isotope data from the Kvitøya Trough supported by AMS-$^{14}C$ dates suggest the older age of the early deglaciation stage in this narrow eastern Svalbard trough. Similar dates were previously published for the St. Anna Trough. However, the deglaciation of local marine-based ice sheet margins continues e.g., within the Franz Josef Land up to date with recent diamicton deposition. The diamicton is commonly overlain by Unit II with a sharp contact in between, deposited during the late deglaciation stage, when the iceberg calving and rafting ceased owing to glacier margins retreat to the land. The Unit II is mainly composed of fine-grained silty clay with thin fine-grained sand layers and lenses. It is often characterized by thin brown and gray lamination. Both laminated and lenticular sedimentary structures suggest pulsating deposition from glacier meltwater suspension flows likely accelerated by sea ice freezing brines cascading. Coarse IRD is rare or absent. Undulating seismic facies occur in the southern Barents Sea resembling sedimentary waves possibly indicating contourite origin of the thick Unit II. The upper boundary of the Unit II with overlying Holocene marine sediments appears to be also diachronous. Since the LGM until the Recent, the deglaciation was fostered by relatively warm surface and subsurface Atlantic water inflow into the Barents Sea through the western and northern deep troughs, respectively. The glacio-eustatic sea-level rise led to Atlantic water penetration beneath the glacier ice resulting in its melting from below thus promoting ice streams activity during early deglaciation. The distribution of iceberg rafted debris and its petrography depended on the provenances drained by ice streams as well as on the icebergs drift routes and their melting sites. Severe sea ice conditions were favorable for sub-ice sedimentation of the fine-grained diamicton matrix in small freshened basins between still glaciated shoals. Short down-slope suspension flows and seasonal brines cascading were main sedimentation mechanisms during the late deglaciation, although more extensive contour currents were possibly active in the southern Barents Sea. According to micropaleontological data, biological productivity was generally low during the deglaciation likely owing to restricted mixing of the subsurface saline Atlantic water and the surface water freshened by glacier melt-water discharge.

**Keywords:** deglaciation, lithostratigraphic units, microfossils, stable isotopes, Atlantic water

## INTRODUCTION

The late Weichselian Barents Sea continental glacier represented the largest grounded marine based ice sheet of the last glacation in the Northern Hemisphere. According to the prevailing view (e.g., Svendsen et al. 2004), the Barents Sea shelf and surrounding archipelagoes were totally covered by an ice sheet united with the mainly land based Scandinavian ice sheet in the south during the Last Glacial Maximum (LGM) (Figure 1). Its boundaries followed the shelf brake in the west and the north. The eastern boundary in the Kara Sea is still uncertain and the maximum ice margin extent in the area was likely not synchronous with the timing of LGM in other margins (Hughes et al. 2016). The marine based ice sheet apparently covered the deep St. Anna Trough and possibly also the Voronin Trough. Further southward, the ice sheet boundary is commonly depicted along the north-eastern slope of the Eastern Novaya Zemlya trench through the Karskie Vorota Strait to the shallow Pechora Sea which emerged during the LGM sea level low-stand and was likely ice free. Behavior of the ice sheet margin in the south-easternmost Barents Sea – northern Pechora Sea during LGM and early deglaciation is still disputable among researchers based on different interpretation of detailed seismic survey, as well as shallow engineering borehole data (Polyak and Mikhailov 1996; Pavlidis and Polyakova 1997; Gataulin et. al. 2001; Krapivner 2006, 2009, and references therein). The boundary between Barents and Scandinavian ice sheets is arbitrarily drawn along the Nordcap – Kola trench. Both sheets had their own ice distribution centers which discharged glacier ice from both sides into the major ice stream flowing westward towards the Bear Island Trough that developed at the beginning of deglaciation (e.g., Siegert and Dowdeswell, 2002; Ottesen et al. 2005, 2008).

This chapter aims to highlight the influence of global and regional paleoceanographic changes on the deglaciation of the marine based Barents ice sheet since the last glacial maximum (LGM) until the onset of marine environment in the Holocene. A special attention is paid to the role of relatively warm Atlantic Water flow into the Barents Sea. However, it should be taken into account that even during the LGM, Atlantic Water seasonally penetrated into the Arctic Ocean via the Fram Strait and further eastward along the margin of Scandinavian-Barents ice sheet (e.g., Ivanova 2009; Xiao et al. 2015; Ivanova et al. 2016; Figure 1). Therefore, the relatively warm Atlantic water affected the configuration of the ice margin already at the onset of LGM and even much earlier, as shown in (Hughes et al. 2016).

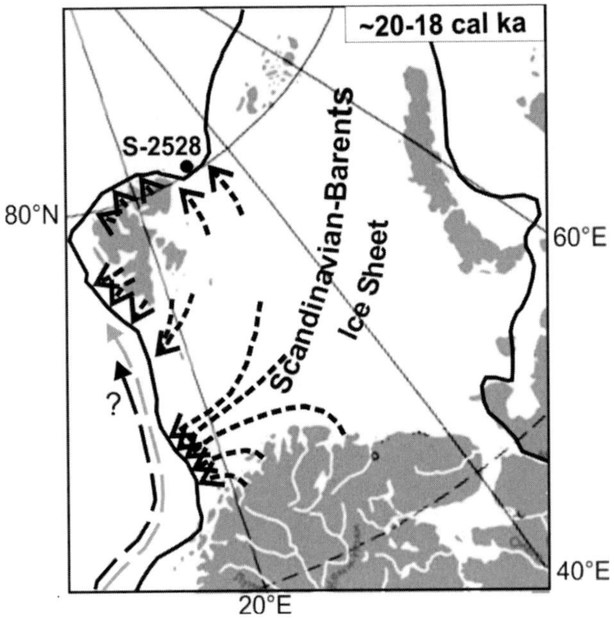


Figure 1. Scandinavian-Barents Sea Ice sheet extend and Atlantic Water flow along its margin at the end of Last Glacial Maximum (LGM), 20-18 cal ka BP (modified from Ivanova et al. 2016). Broken black and gray lines with arrows designate the inflow of surface and subsurface Atlantic Water, respectively. Ice streams are shown schematically by black dashed lines with arrows after (Landvik et al. 1998; Ottesen et al. 2005, 2008; Fransner et al. (in prep.)).

Here, we consider the deglaciation of the marine-based area of the Barents ice sheet. This process differs from the decay of land based ice sheets in dramatic influence of oceanographic parameters (temperature, salinity, stratification of water masses, their lateral and vertical dynamics, mixing with glacier melt water, global sea level change, etc.). These parameters can be reconstructed by proxy records preserved in glaciomarine sediment successions recovered by sediment cores and boreholes. Unfortunately, the scarcity of calcareous fossils (planktic and benthic foraminifers, mollusk shells) in the glaciomarine sediments of the Barents Sea strongly limits their dating and thus compilation of age models especially for the early deglaciation phase. Only a few datings are considered from the marine based Barents-Kara ice sheet in the comprehensive summary by Hughes et al. (2016). This led us to focus more attention on the lithostratigraphic approach in the present study, along with recent new radiocarbon datings and more adequate paleoceanographic interpretation of the data obtained.

## STRATIGRAPHIC FRAME

Postglacial glaciomarine deposits of the Barents Sea shelf recovered by numerous gravity cores from shelf depressions we subdivide into two distinct lithostratigrapic units corresponding to early (Unit III) and late (Unit II) deglaciation phases (e.g., Ivanova et al. 2002; Murdmaa et al. 2006). A similar subdivision, but with somewhat different nomenclature, is used by most other researchers (e.g., Lubinski et al. 1996, 2001; Polyak et al. 1995, 1997). The glaciomarine deposits are overlain by Holocene marine sediments with a minor contribution of glacial-derived material (Unit I) not considered in this chapter. The lithostratigraphic succession of the deglaciation is tentatively correlated with the standard Northern Hemisphere stratigraphic scale, sea level and NGRIP climatic records (Figure 2). However, the diachronous character of unit boundaries should be taken into account. Examples of selected core sections demonstrating the deglacial units succession are shown for the northern (Figure 3), central and southern (Figures 4-6) Barents Sea.

According to the hypothesis by Clark et al. (2009), the onset of the last deglaciation in the Northern Hemisphere was marked by a series of meltwater events beginning at c.19 cal ka BP.

The start of deglaciation in the Fram Strait and Barents Sea area including the continental margin and slope of Spitsbergen is inferred from a rapid lightening of planktic oxygen isotope values ($\delta^{18}O$) and a contemporaneous minimum in carbon isotopes ($\delta^{13}C$) (Jones, Keigwin 1988; Hebbeln et al. 1994, Stein et al. 1994; Knies et al. 1998, 2000; Ivanova et al. under review; see also Figure 5 in this chapter). For several millennia after the LGM, the Barents Sea ice sheet is known to be drained by ice streams and ice flows, especially active in the troughs, under the control of sea level rise (e.g., Landvik et al. 1998; Siegert and Dowdeswell, 2002; Ottensen et al. 2005, 2008). In the troughs, gradual ice retreat was punctuated by minor re-advances or still-stands and progresses also through iceberg calving (e.g., Hogan et al. 2010a,b).

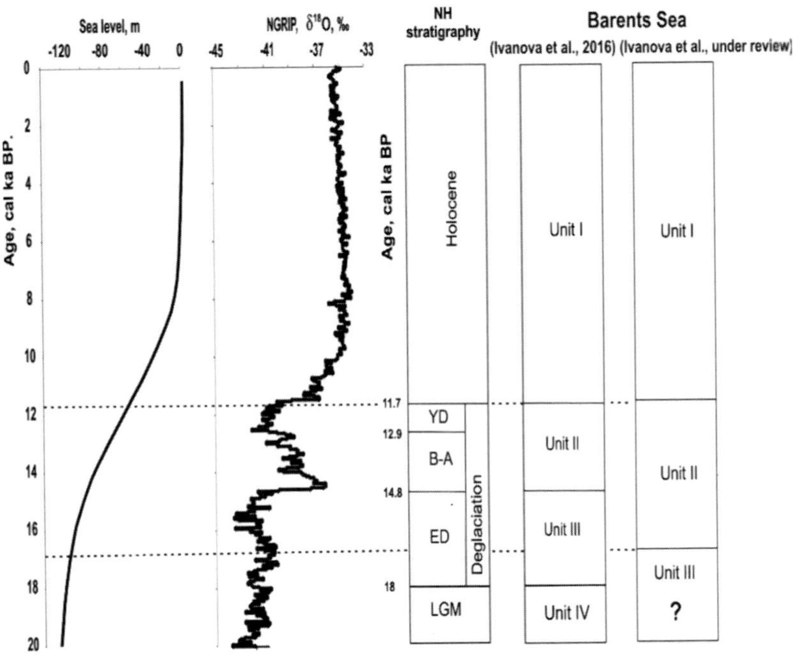


Figure 2. Correlation of lithotostratigraphic frame for the Barents Sea with the Northern Hemisphere (NH) stratigraphy, sea level record (Fairbanks 1989) and NGRIP climatic record (North Greenland Ice Core Project members 2004) for the last 20 ka (after Ivanova et al. under review with additions). Dotted lines designate correlation of Unit II in the NW Barents Sea with synchronous intervals of the other records. Note the diachronous units II/III boundary in the Kvitøya Trough (right column) comparing to the other parts of the Barents Sea and an uncertainty with the lower boundary of Unit III. LGM - Last Glacial Maximum; ED - early deglaciation; B–A - Bølling–Allerød; YD - Younger Dryas.

## THE EARLY DEGLACIATION

The paleoenvironment during the early deglaciation phase is reconstructed using studies of the Unit III represented by gray (commonly dark gray) glaciomarine diamicton, a disordered mixture of unsorted coarse-grained terrigenous debris (from sand to pebble- and boulder-size rock fragments) with very fine-grained matrix commonly dominated by the size fraction <0.001 mm. The bimodal grain size distribution indicates different deposition modes of two main diamicton components. The wide variability in size, petrography, and rounding state of rock fragments suggests iceberg rafting of the coarse fractions (Figure 3; Murdmaa et al. 2006; Ivanova et al. 2016). Therefore, it supports the idea of iceberg calving as an important factor of the ice sheet destruction during the early deglaciation phase (Hogan et al. 2010a,b). However, glacier proximal debris flows also contribute to the diamicton deposition in shelf depressions (e.g., Lubinski et al. 2001). The fine-grained matrix is apparently derived from glacier melt-water plumes (so-called glacial milk) and settled in calm sub-ice conditions in the newly formed considerably freshened marine basins under almost permanent sea ice cover (Murdmaa et al. 2006). Based on the grain size distribution and clay mineral composition (high kaolinite content not characteristic for soils from high latitudes), we hypothesize that a considerable proportion of the diamicton matrix was derived from the remote aeolian dust stored in the glacier ice and released during its melting at the early deglaciation phase (Murdmaa et al. 2006).

We propose a glaciomarine origin of the diamicton based on rare, but persistent occurrence of well-preserved planktic and benthic foraminiferal tests (in some samples also pteropod shells) belonging to extant species, grain size distribution of the diamicton matrix dominated by the finest clay fraction (<0.001mm), salinity of interstitial water close to considerably freshened seawater (Murdmaa et al. 2006).

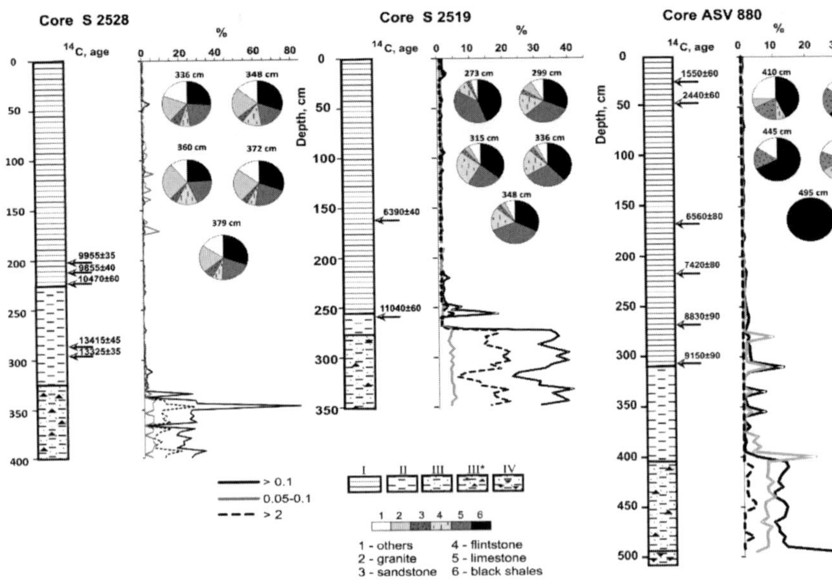


Figure 3. Cores S-2519 and S-2528 from the eastern Svalbard troughs and core ASV-880 from the Franz Victoria Trough: lithology, AMS-14C dates, grain size distribution and petrography of IRD from selected diamicton samples (modified from Ivanova et al. 2016 and Ivanova et al. under review). Variability of rock types in the iceberg rafted coarse debris contrasts with only single black shale fragment in the glacier till (Unit IV) sample from core ASV-880 (495 cm). Unit II is homogeneous in core S-2528; it demonstrates variegated lamination in core S-2519, and contains thin lenses of fine-grained sand intercalated with clay in core ASV-880. For the core locations see Figure 8.

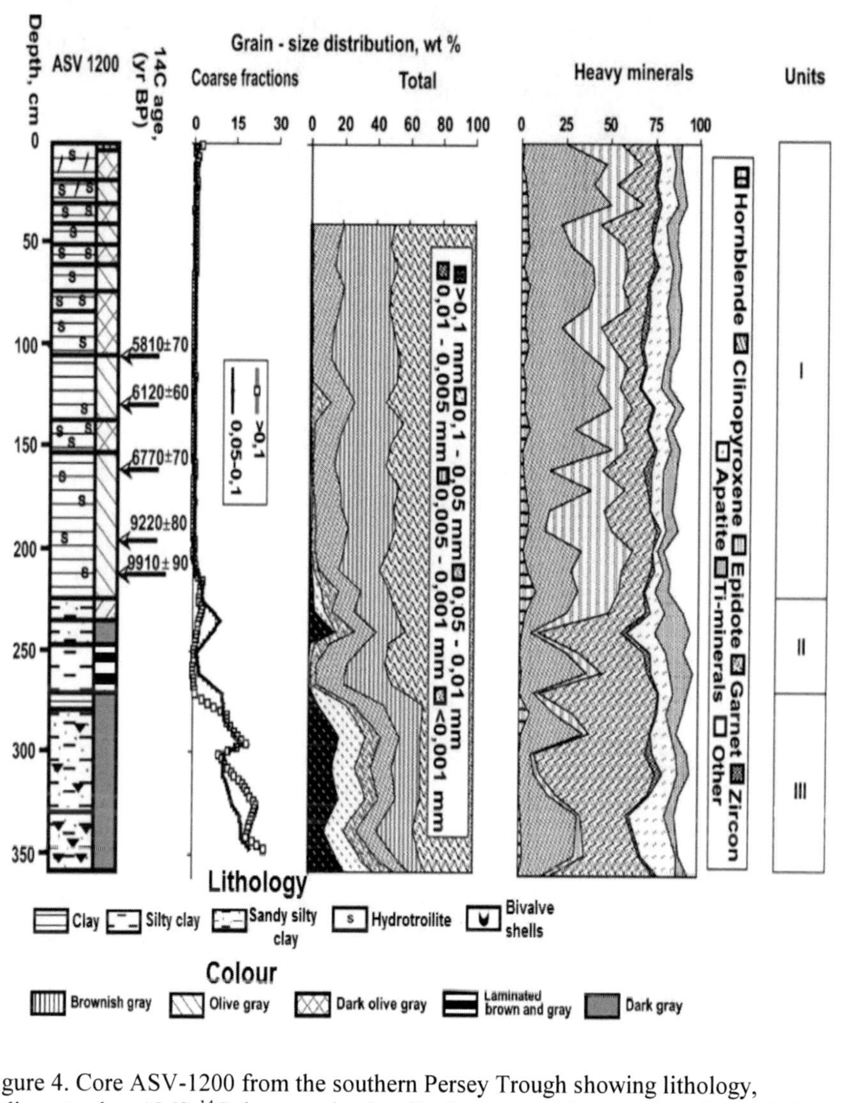


Figure 4. Core ASV-1200 from the southern Persey Trough showing lithology, sediment color, AMS-$^{14}$C dates, grain size distribution, and heavy mineral associations in units III, II and I (modified from Murdmaa et al. 2006). Note the color change and sharp decrease in coarse fraction (IRD) content at units III/II boundary; change in heavy minerals association at units II/I boundary marking the transition from distal glaciomarine to the Holocene marine environment. The coarser grained gray interbed in Unit II suggests a re-advance event of the glacier margin. For the core locations see Figure 8.

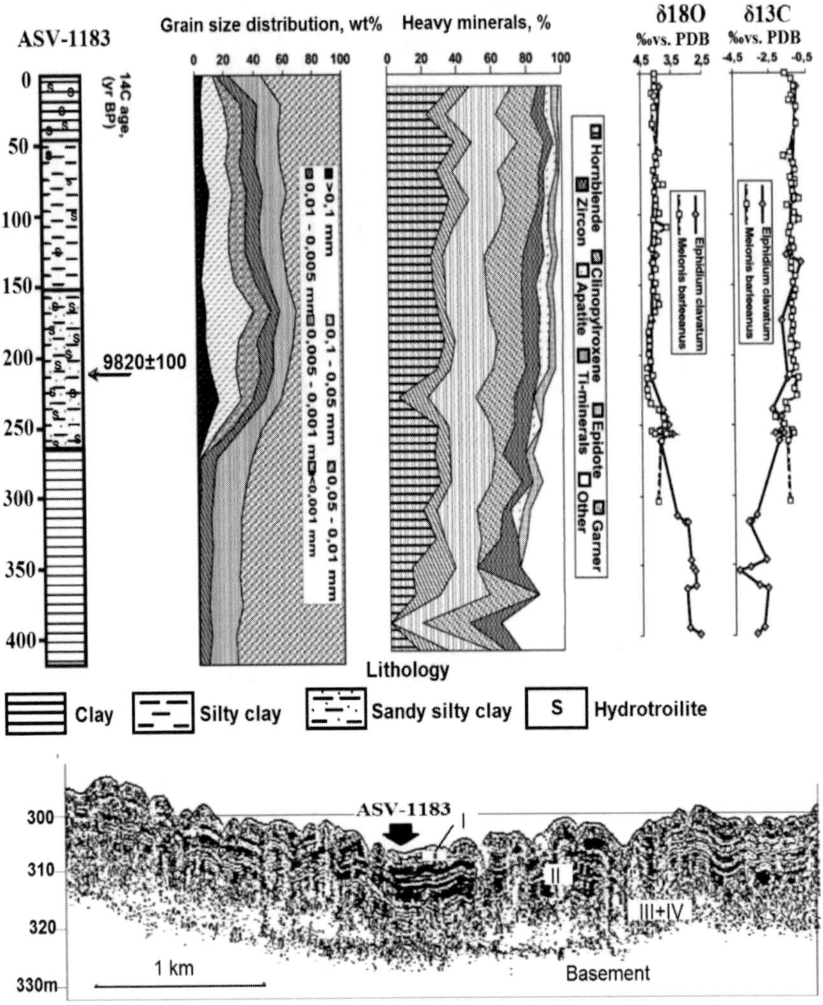


Figure 5. Core ASV-1183 from the southern Central Deep showing thick Unit II represented by very fine-grained grayish brown clay strongly dominated by <0.001 mm fraction. Its heavy mineral association is similar to that of Unit I suggesting a common provenance. Light $\delta^{18}O$ and $\delta^{13}C$ values in the late deglaciation interval are interpreted as glacier melt-water and/or riverine freshwater signals. The *Parasound* seismic profile below demonstrates undulating 20 m thick Unit II overlying opaque diamicton of Units III+IV (modified from Murdmaa et al. 2006). For the core locations see Figure 8.

The diamicton unit constitutes the basal layer in numerous cores with an apparent thickness up to 3 m. Core ASV-880 from the Franz Victoria Trough recovered below the diamicton a very stiff dark gray deposit consisting of angular black (bituminous) shale fragments up to 12 cm in size (Figure 3, Unit IV). High density, very low water content (18%) and strongly freshened interstitial water salinity (13‰) suggest that this core catcher sample represents the top of basal moraine (subglacial till) deposited during the LGM, before the beginning of early deglaciation (Murdmaa et al. 2006). Many boreholes from the south-eastern Barents Sea recovered Mesozoic bedrocks beneath a thick (tens of meters) diamicton layer. However, in several boreholes, a sharp increase in density of the diamicton suggests deposits of an earlier glaciation separated from the last one by a long-term hiatus well reflected in seismic records (Krapivner 2006).

At the early deglaciation, severe cold conditions with dense perennial or quasi-perennial sea-ice cover, low bottom-water salinity and low bioproductivity are inferred from the scarce occurrence of *in situ* benthic foraminifers and dinocysts, low (of most likely terrigenous) TOC content and the absence of planktic foraminifers. However, the rare specimens of indicative benthic foraminiferal species (*M. barleeanus, C. lobatulus* and others) found in the cores from Kvitøya and Franz Victoria troughs might reflect the very early subsurface Atlantic Water influence (Ivanova 2009; Ivanova et al. under review).

Several studies proposed Atlantic Water flow along the Svalbard-Barents Sea continental slope since the LGM and possible formation of coastal polynyas caused by katabatic winds (Xiao et al. 2015 and references therein; Falardeau et al. under review). The Atlantic Water may have fostered the calving of the ice sheet margins and sliding of the ice on the Barents Sea floor, thus contributing to accelerate its decay, notably through ice streams and flows (Landvik et al. 1998; Siegert and Dowdeswell 2002; Svendsen et al. 2004; Ottesen et al. 2005, 2008; Dowdeswell et al. 2010; Hogan et al. 2010a, b; Ivanova et al. 2016; Hughes et al. 2016; Fransner et al. in prep.).

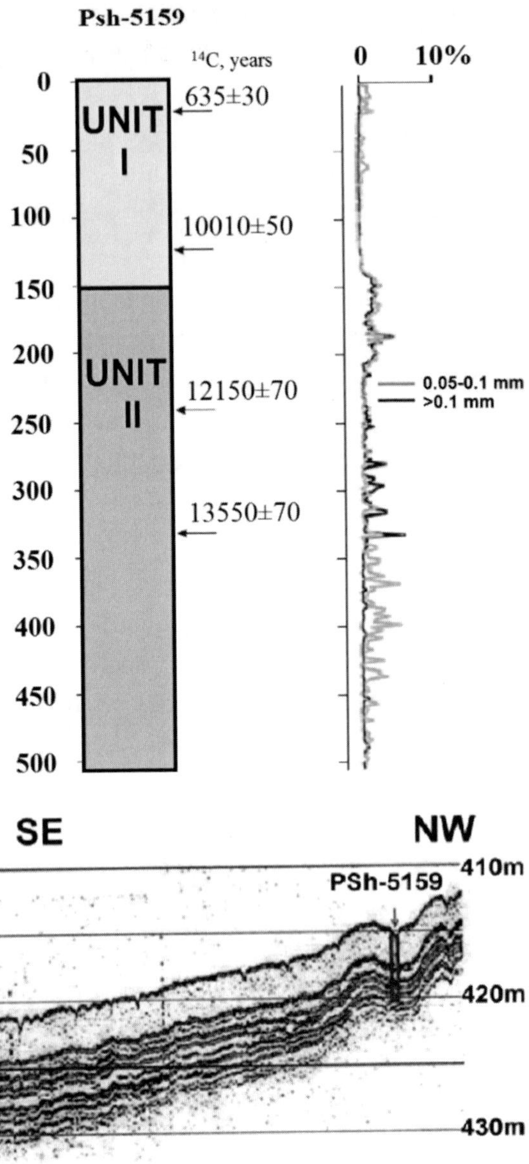


Figure 6. Core PSh-5159R from the Ingøydjupet Depression showing generalized thick Unit II represented by grayish brown clay and silty clay intercalated with numerous thin (millimeter-scale) lenses of fine-grained sand. AMS-$^{14}$C dates indicate its high sedimentation rates. High-resolution *SES-2000 medium* seismic profile (below) demonstrates distinctly stratified Unit II undulating (as sedimentary waves) near the site PSh-5159R. For the core locations see Figure 8.

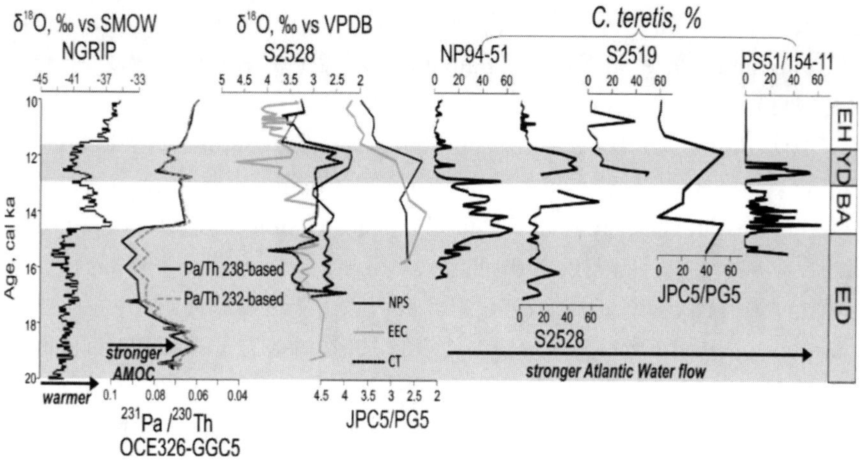


Figure 7. Correlation of proxy time series from Greenland (North Greenland Ice Core Project members 2004), core OCE326-GGC5 (McManus et al. 2004) from the Atlantic used as the reference record of Atlantic Meridional Overturning Circulation (AMOC), Kvitoya and Erik Eriksen troughs (cores S2528 and S2519, Ivanova et al. under review), Franz Victoria Trough (core JPC5/PG5, Lubinski et al. 2001), northern Svalbard margin (core NP94-51, Ślubowska et al. 2005); and Laptev Sea (core PS51/154-11, Taldenkova et al. 2012) over the deglaciation (modified from Ivanova et al. under review). See Figs. 2 and 8 for stratigraphic abbreviations and core locations, respectively. Oxygen isotopes measured on planktic species NPS – *Neogloboquadrina pachyderma* sin., and benthic infaunal species: EEC – *Elphidium excavatum* forma *clavata* and CT – *Cassidulina teretis*.

Recent multi-proxy reconstructions (e.g., Chauhan et al. 2016; Ivanova et al. 2016; Ivanova et al. under review) suggest significant Atlantic Water influence in the Barents Sea troughs and margins during the Heinrich stadial 1 (HS1) when the AMOC was weaken as compared to the Bølling–Allerød (B-A) and Holocene (McManus et al. 2004; Figure 7). The increased abundance of *Cassidulina teretis* and planktic foraminiferal occurrences within the HS1 interval concurrently indicate an enhanced Atlantic Water flow into the Kvitøya, Franz Victoria and St. Anna troughs (Lubinski et al. 2001; Ivanova et al. under review; Figure 7), on the Svalbard northern slope (Ślubowska et al. 2005) and continental margin (Chauhan et al. 2016), and to the western Barents Sea (Rasmussen et al. 2007). The subsurface Atlantic Water likely enters the Nordic Seas and Barents Sea under the sea-ice cover and/or strong pycnocline in summer.

In such conditions, the heat reservoir may build up in subsurface layer (e.g., Rasmussen et al. 2014; Chauhan et al. 2016) fostering the ice sheet decay and iceberg calving.

The timing of transition from the proximal to distal glaciomarine conditions is yet poorly constraint across the major part of the basin. The Barents Sea ice sheet retreated from the deep troughs and shelf depressions likely 18 – 15 cal. ka BP, and glaciers only retained on archipelagoes and shoals (Vorren and Laberg 1996; Polyak et al. 1997; Elverhøi et al. 1998; Landvik et al. 1998; Knies et al. 1998; Hald et al. 1999; Lubinski et al. 2001; Hughes et al. 2016; Ivanova et al. under review), thus opening the marine passages. However, Lubinski with co-authors (2001) argue that the ice sheet decay started first in the southern Barents Sea. Most researchers believe that the major part of the basin was already ice-free at 15-14 cal. ka BP (Figure 8a; e.g., Klitgaard-Kristensen et al. 2013; Ivanova et al. 2016; Hughes et al. 2016).

## THE LATE DEGLACIATION

The late deglaciation phase (Unit II) was characterized by a distal glaciomarine environment. At that time, the glacier front retreated inland from the Novaya Zemlya coast (Forman et al. 1999; Zeeberg et al. 2001). Gataullin and co-authors (2001) argue that it happened after the Younger Dryas (YD). Several studies suggested that remains of the Barents Sea ice sheet were melted on shoals and southern Novaya Zemlya islands at the beginning of Holocene (e.g., Vorren and Laberg 1996; Landvik et al. 1998). Retreat of the glacier front from the coastline resulted in almost total cessation of the iceberg rafting of coarse debris in the open Barents Sea (Murdmaa et al. 2006). Settling of fine-grained suspended material derived from glacier meltwater became the main sedimentation mechanism. Bioproductivity was generally low owing to severe sea ice conditions (Murdmaa and Ivanova 1999; Ivanova et al. 2002).

In most gravity cores and boreholes throughout the Barents Sea, Unit II is separated from the underlying Unit III by a sharp contact that expresses significant changes in sedimentation processes or even an erosion hiatus. The ice-rafted coarse debris abundant in Unit III almost disappears indicating cessation of iceberg calving (Figures 3, 4). A more or less distinct color change commonly characterizes the units boundary (Figure 4). Instead of uniform gray diamicton, there appear yellowish or brownish color shades in the fine-grained Unit II sediments suggesting suboxic early diagenesis, although greenish gray (reduced) varieties also occur (e.g., core ASV-880 from the eastern marginal trench of the Franz Victoria Trough, Figure 3). Up to several decimeter thick intervals of distinct thin lamination with alternating millimeter-scale brown and gray laminae are observed in several cores from the shelf depressions. These variegated laminae were likely deposited by pulsating melt-water suspension flows from neighbor shoals or islands where remains of melting ice sheet have survived. A variegated laminated sequence is described in Unit II of core S-2519 from a local hole in the Erik Erikson Trough (Figure 3) where a polynya likely existed.

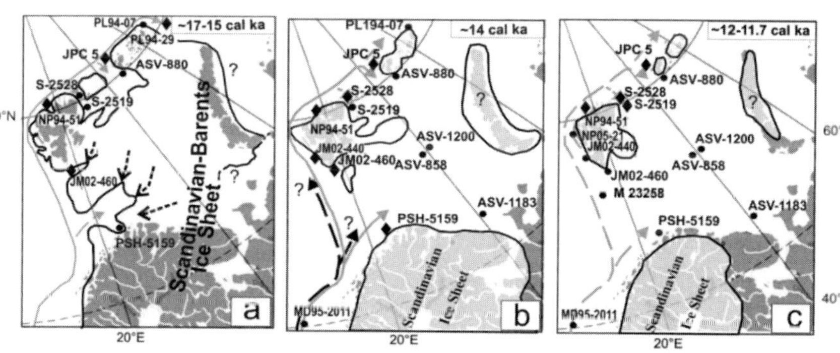


Figure 8. Schematic maps of the Atlantic Water inflow and deglaciation in the Barents Sea area for different time intervals (cal ka BP): (a) 17–15 (ED), (b) ~14 (B–A), (c) ~12.0–11.7 (YD), accounting for original data and reconstructions from several publications and modeling results (modified from Ivanova et al. 2016, see list of references in the aforementioned publication). The solid black line shows boundaries of glaciers. Black diamonds/dots with core numbers show stations with a defined strong/weak "signal" of subsurface Atlantic Water . Solid black and gray arrows designate the inflow of surface and subsurface Atlantic Water, respectively, into the Barents Sea; dashed arrows mark the weak inflow. For abbreviations, see captions to Figure 2.

Core ASV-1200 recovered a succession of three lithofacies (Figure 4) which reflects paleonvironmental development of the Persey Trough during the late deglaciation. The dark gray diamicton of Unit III is overlain by laminated interval suggesting pulsating deposition from short suspension flows. A stronger debris flow followed being possibly induced by an ice cap re-advance on the adjacent rise, or on the contrary, by more intense melting related to climate warming. A graded sandy silty clay bed above. likely related to the same debris flow. coincides with the transition to the Holocene marine Unit I. The transition is also marked by changes in heavy mineral association which documents appearance of provenances different from those inferred for deglacial sediments mineralogy.

Core ASV-1183 from the southern margin of the Central deep (depth 330 m) recovered homogeneous grayish brown very fine-grained (60 – 75% of < 0.001 mm fraction) clay beneath olive gray Holocene marine sediments (Figure 5). A succession of color-banded fine-grained mud ~20 m thick, overlies gray diamicton in a nearby borehole (Krapivner 2009), thus representing an analogue of the Unit II recorded by the *Parasound* seismic profile (Figure 5).

The more or less oxygenated bottom water environment inferred from the Unit II sediment color likely indicates oxygen-rich surface waters descending either as sea ice freezing brines cascading or a low density suspension gravity flows from shoals. Sedimentary structures with variegated lamination or thin sand intercalations suggest lateral suspension flow deposits and thus pulsating sedimentation mode instead of continuous accumulation from the water column.

Thin well sorted fine sand to coarse silt lenses and laminae abundant in the Unit II clayey mud from the Ingøydjupet Depression (Figure 6) and found also in many other cores from deep shelf depressions support the assumption about activity of low-density suspension flows during the late deglaciation. High-resolution *Parasound* and *SES-2000 medium* seismic profiling revealed meandering erosion channels in the Persey Deep and in other shelf depressions. Distinct undulation of seismic reflectors near the ASV-1183 site in the southeastern Barents Sea (Figure 5) may well represent small meter-scale sedimentary waves. Wavy acoustic structures are also recorded in stratified seismic facies from the Ingøydjupet Depression where rapidly accumulated Unit II mud (estimated sedimentation rate of 65 cm/ka) with numerous well sorted fine-grained sand lenses and laminae are recovered by core PSh-5159 (Figure 6) beneath the acoustically transparent Holocene deposits (estimated sedimentation rate of 11 cm/ka). In both regions, the undulated seismic facies likely evidence for a bottom currents (or distal gravity driven suspension flows) activity and abundant fine-grained terrigeneous material supply from the land-based melting glaciers. The contribution of paleo-Pechora River discharge into the glacier ice-free basin in the SE Barents Sea during the late deglaciation phase is also discussed by some researchers (e.g., Krapivner 2009). Absence of any considerable change in the heavy minerals association at the units II/I boundary in core ASV-1183 indicating a common provenance, in spite of a dramatic difference in sediment types (Figure 5), seems to support this suggestion.

Therefore, sedimentation during the late deglaciation phase was controlled by fine-grained suspended material supply derived both from melting remnants of the ice sheet and reworking of loose diamicton matrix from the submarine topographic highs and slopes. These main sediment sources had together formed a near-bottom nepheloid layer which might act as heavy liquid able to stimulate low-density gravity flows over the unusually rough subglacial bottom relief of the Barents Sea shelf. Intensity of the suspension flows was accelerated by cascading of sea ice freezing brines when perennial sea ice cover was replaced by seasonal one (e.g., during the B-A interstadial).

The northern part of the Barents Sea was perhaps permanently covered by sea ice, both locally formed and drifted from the Arctic Ocean during the late deglaciation. Polar surface water, in spite of an intense input of subsurface Atlantic Water during the B-A interstadial, was also noted in cores NP94-51 and JM02-440 taken off the northern and western Spitsbergen coast respectively (Ślubowska et al. 2005, 2007). The surface waters seem to be also cold and seasonally covered by sea ice in the southwestern Barents Sea even during the B-A warming, as it follows from the extremely scarce occurrence of planktic foraminifers in core PSh-5159R (Chistyakova et al. 2010).

The pulsating sedimentation likely created an unfavorable environment for both macro- and microbenthos (Murdmaa and Ivanova 1999; Ivanova and Murdmaa 2001) in the major part of the basin. This suggestion is consistent with existing data on the seasonal development of benthic foraminifers near the modern glaciers front (Korsun and Hald 2000). Rather rare occurrences of extant foraminiferal species may also be related to dissolution of their calcareous tests. Sea ice freezing brines and nepheloid flows carry dissolved oxygen and carbon dioxide from the sea surface to the bottom of troughs, thus promoting dissolution of calcareous microfossils (Steinsund and Hald 1994; Hald and Steinsund 1996).

An intense glacier melting during the late deglaciation phase was amplified by warm Atlantic Water inflow into the Barents Sea via the deep troughs (Figure 8b,c). Available evidence from the northern troughs and margins suggest increased subsurface Atlantic Water inflow during the B-A warming (Klitgaard-Kristensen et al. 2013), at the end of YD-Preboreal (Lubinski et al. 2001), and during early-to-middle Holocene (e.g., Duplessy et al. 2001, 2005; Ślubowska et al. 2005). In particular, a pronounced increase in the Atlantic Water flow and/or warming of this water mass in the bottom layer was recently inferred from the high relative abundance of *C. teretis* (30-66%) in core S-2528 at *ca.* 13.6 cal ka BP, i.e., during the Allerød interstadial (Ivanova et al. under review; Figure 7). In the same core, an increased percentage of dinocyst taxa often associated with Atlantic Water suggests at least episodical influence of this water mass in the surface layer. Planktic and benthic foraminifers, notably *C. teretis*, also occur in the laminated B-A sediments of core S-2519 thus implying the Atlantic Water influx to Erik Eriksen Trough (Ivanova et al. under review). This is in line with the previous studies documenting high abundance of *C. teretis* at B-A from the locations southward of the Kvitøya Island (Klitgaard-Kristensen et al. 2013), from the northern troughs (Lubinski et al. 2001); along the Barents Sea margin (Ivanova et al. 2016 and references therein), Yermak Plateau (Wollenburg et al. 2004) and Laptev Sea (Taldenkova et al. 2012) (Figure 7). Lubinski and co-authors

(2001) suggested that stronger Atlantic Water influx was facilitated by isostatic submergence of the Northern Barents Sea troughs.

In the western and southwestern parts of the Barents Sea, surface and subsurface Atlantic Water flows are mainly documented to be enhanced at about the same time intervals (e.g., Sarnthein et al. 2003; Husum and Hald 2004; Rasmussen et al. 2007, 2012; Aagaard-Sørensen et al. 2010; Chistyakova et al. 2010; Risebrobakken et al. 2010, 2011; Berben et al. 2014). On the contrary, the diverse B-A benthic assemblages of the Ingøydjupet Depression were dominated by the Atlantic-affiliated species, whereas the percentage of the typical Arctic shelf species, notably *Elphidium clavatum*, increases through the YD (Chistyakova et al. 2010). High abundance of warm-water benthic foraminifers indicates the persistent Atlantic Water presence in the bottom layer of the southwestern Barents Sea whereas the scarcity of planktic foraminifers most likely points to very low sea-surface temperatures and seasonal ice cover.

The resumption of deep-water formation in the North Atlantic (Figure 7; McManus et al. 2004) during the broad B-A warming in the North Atlantic area (Rasmussen and Thomsen 2008) fostered an increased subsurface Atlantic Water flow to the western and southwestern Barents Sea and further to the Arctic (Ślubowska et al. 2005; Chistyakova et al. 2010; Aagaard-Sørensen et al. 2010; Chauhan et al. 2016; Ivanova et al. 2016).

Meanwhile, the presence of Atlantic-affiliated planktic and benthic foraminiferal species in the Unit II sediments suggests episodic Atlantic Water penetration also to the Central Barents Sea via the troughs (Murdmaa and Ivanova 1999; Ivanova and Murdmaa 2001). This suggestion is supported by our data from cores ASV-1183, PSh-5147, and ASV-1200, as well as by other studies (Polyak and Solheim 1994; Lubinski et al. 1996, 2001; Hald et al. 1999) about thin interbeds in Unit II sediments with relatively abundant benthic foraminifers including species characteristic of Atlantic Water.

In the core S-2528 from the Kvitoya Trough, generally small differences in the $\delta^{18}O$ of planktic and benthic foraminiferal values suggest subsurface-to-bottom waters mixing during the B-A interval, below a relatively well stratified surface layer (Figure 7; Ivanova et al. under review). The surface waters were receiving the meltwater plumes from the retrieving glaciers likely from the Svalbard margin (e.g., Klitgaard-Kristensen et al. 2013; Chauhan et al. 2016). Low surface productivity is inferred from low concentrations of biogenic remains and low TOC content in the cores studied from the area (e.g., Ivanova et al. under review).

However, in most of the Barents Sea area, more favorable conditions for production and/or preservation of benthic foraminiferal fauna appeared only episodically (e.g., Ivanova 2009). The basin became perhaps partially ice-free for short periods during the late deglaciation phase, possibly due to polynya formation where high productivity developed. Hence, the deglaciation was characterized by unstable paleoceanographic conditions. The basin was filled with considerably freshened seawater due to glacier meltwater delivery, as well as river water inflow via White, Pechora, and Kara seas, as evidenced by decreased interstitial water salinity in Unit II sediments (Murdmaa et al. 2006). The strong freshening is also confirmed by depleted oxygen and carbon isotope values of benthic foraminifer species *Elphidium clavatum* from core ASV-1183 collected in the southern part of the Central Deep (Figure 4) (Ivanova et al. 2002; Murdmaa et al. 2006). The freshwater isotope signal was likely transferred to the bottom water by brines formed over nearby shoals during the winter sea ice freezing (Bauch and Bauch 2001; Duplessy et al. 2001; Hillaire-Marcel and de Vernal, 2008).

A meltwater isotope signal is also noted at 12.8 and 11.4 cal. ka BP in the northern Barents Sea (Lubinski et al. 2001), at ca. 13 and 12 cal. ka on the eastern slope of the Laptev Sea (Spielhagen et al. 2005), and as short episodes in several locations of the Arctic Ocean within the interval 18.8 – 13 cal. ka BP (Stein et al. 1994; Nørgaard-Pedersen et al. 1998; Poore et al. 1999). Some authors assumed that the freshwater signal reflects not only the Barents Sea ice sheet melting, but also an increased influence of the Ob, Yenisei, and other Siberian rivers runoff due to their seaward propagation prior to the YD and at the YD/Preboreal transition (Polyak et al. 2000; Lubinski et al. 2001).

The low planktic $\delta^{18}O$ values during the YD (Figure 7) reflect dilution with freshwater, probably as the result of freshwater/meltwater discharge from the Arctic related to the northward drainage of paleo-lake Agassiz through the McKenzie pathway (Tarasov and Peltier, 2005; Murton et al. 2010; Hillaire-Marcel et al. 2013). Besides, the abovementioned mechanism of transfer of isotopically light waters from the surface to the subsurface and bottom water layers through the release of brines should be taken into account. Cold conditions and regional sea ice formation are documented by the dinocyst assemblages characterized by the almost exclusive occurrence of heterotrophic taxa associated with dense, possibly sub-perennial sea ice (Ivanova et al. under review). In general, cold conditions prevailed during this interval over the Nordic and Barents Seas (Eldevik et al. 2014 and references therein). The peak of low NPS $\delta^{18}O$ values in core S-2528 (Figure 7) seems to be rather synchronous with similar peaks in cores JPC5 from the Franz Victoria Trough (Lubinski et al. 2001), PL94-67/C from St. Anna Trough (Hald et al. 1999) and PS2458 from the Laptev Sea (Spielhagen et al. 2005) which were interpreted as resulting from extensive iceberg discharge and/or meltwater pulse.

During the YD, high percentages of *C. teretis*, the occurrence of planktic and diverse benthic foraminiferal assemblages characterize cores S-2528 and S-2519 (Figure 7), thus suggesting a substantial subsurface Atlantic Water flow into the Kvitøya – Erik Eriksen troughs most likely from the Fram Strait and along the Eurasian margin (Figure 8c; Ivanova et al. 2016). This finding is in line with benthic foraminiferal data from the Yermak Plateau (Wollenburg et al. 2004), Franz Victoria Trough (Lubinski et al. 2001) and Laptev Sea (Taldenkova et al. 2012)(Figure 7). In spite of the reduced strength of overturning circulation during the YD cooling (Figure 7; McManus et al. 2004) the submerging northward Atlantic Water flow probably occurred during this interval like during the HS1 (Ivanova et al. 2016). At the YD/Preboreal transition, the reorganization of ocean circulation was appearently accompanied by a strong Atlantic Water inflow to the Nordic Seas (Risebrobakken et al. 2011) and to the Eurasian margin (Ivanova et al. 2016; Ivanova et al. under review).

Considering the deglaciation, one has to take into account changes in the bathymetry of the basin owing to glacio-isostatic uplift of the area and glacio-eustatic sea level rise (Fairbanks et al. 1989; Lambeck 1995; Kaplin and Selivanov 1999). The uplift was much faster in the southern Barents Sea. The northern shelf was by 200 m deeper than today during the late deglaciation phase (15 – 11.4 cal. ka BP), whereas depths of the southern Barents Sea were close to present values. Depth of the Bear Island Trough reached 400 m, and Atlantic waters might easily penetrate into the Barents Sea. Even at the beginning of the Holocene, the northern shelf was still by 100 m deeper than today (Vorren and Laberg 1996; Forman et al. 1997; Lubinski et al. 2001).

## Conclusion

Reconstructed paleoceanographic environments of early and late deglaciation phases principally differ from each other. Ice streams promoted by Atlantic Water penetration beneath the marine based grounded ice sheet, iceberg calving, and coarse debris rafting resulting in glaciomarine diamicton deposition in newly formed freshened basins characterize the early deglaciation. Onset of the late deglaciation phase was marked by the ice sheet margin retreat to the land that led to cessation of iceberg calving and rafting. Gravity driven melt-water suspension flows became the major sedimentation process. Biological productivity increased locally due to intense Atlantic Water inflow, but was still generally low, likely because of strong surface water stratification. Pulsating sedimentation inhibited the development of benthic communities.

The available data document diachronous deglaciation in the Barents Sea with iceberg calving ending at ~17 cal ka BP in the northwest (Kvitoya Trough and likely northern Franz Victoria Trough) and later on in the other parts of the basin. Multi-proxy records also indicate penetration of a subsurface Atlantic Water in the northern and western troughs since ~17 cal ka BP. This water flowed through the Fram Strait, then along the Eurasian margin into the NW Barents Sea and further into the Laptev Sea, peaking during B-A and YD/Preboreal times. However, a dense sea-ice cover with polynyas at some locations likely persisted in the Barents Sea through the deglaciation.

## ACKNOWLEDGMENTS

We greatly appreciate stimulating discussions with Jean-Claude Duplessy, Leonid Polyak, Bjørg Risobrobakken, Riko Noomets, Anne de Vernal and Robert Spielhagen. We thank Nikita Simagin for the assistance in figures and references preparing. We owe our sincere thanks to Thomas Cronin for improving the English and to Yelena Polyakova for critical reading of the Chapter. This study is a contribution to the project 14-50-00095 funded by the Russian Science Foundation.

## REFERENCES

Aagaard-Sørensen S., Husum, K., Hald, M., Knies, J. Paleoceanographic development in the SW Barents Sea during the Late Weichselian-Early Holocene transition. *Quat. Sci. Rev.* 2010, *29*, 1–15.

Bauch, D., Bauch H. A. Last glacial benthic foraminiferal δ18O anomalies in the polar North Atlantic: A modern analogue evaluation. *J. Geophys. Res.* 2010, *106(5)*, 9135-9143.

Berben, S. M. P., Husum, K., Cabedo-Sanz, P., Belt, S. T. Holocene sub-centennial evolution of Atlantic water inflow and sea ice distribution in the western Barents Sea. *Clim. Past*. 2014, *10*, 181–198.

Chauhan, T., Rasmussen, T. L., Noormets, R. Palaeoceanography of the Barents Sea continental margin, north of Nordaustlandet, Svalbard, during the last 74 ka. *Boreas*, 2016, *45*, 76–99.

Chistyakova, N., Ivanova, E., Risebrobakken, B., Ovsepyan, E., Ovsepyan, Ya. Reconstruction of Postglacial environments in the southwestern Barents Sea. *Oceanology*. 2010, *50(4)*, 573–581 (English translation).

Clark, P. U., Dyke, A. S., Shakun, J. D., Carlson, A. E., Clark, J., Wohlfarth, B., Mitrovica, J. X., Hostetler, S. W., McCabe, A. M. The last glacial maximum. *Science* 2009, *325*, 710–714.

Dowdeswell J. A., Hogan, K. A., Evans, J., Noormets, R., Cafaigh C. O., Ottersen, D. Past ice-sheet flow east of Svalbard interfered from streamlined subglaciar landforms. *Geology*, 2010, *38, 2*, 163-166.

Duplessy J. C., Ivanova E. V., Murdmaa I. O., Paterne M., Labeyrie L. Holocene paleoceanography of the Northern Barents Sea and variations of the northward heat transport by the Atlantic Ocean. *Boreas* 2001, *30(1)*, 2-16.

Duplessy J. C., Cortijo E., Ivanova E. V., Khusid T., Labeyrie L., Levitan M., Murdmaa I. O., Paterne M. Paleoceanography of the Barents Sea during the Holocene. *Paleoceanography*, 2005, *20*.

Eldevik, T., Risebrobakken, B., Bjune, A. E., Andersson, C., Birks, H. J. B., Dokken, T. M., Drange, H., Glessmer, M. S., Li, C., Nilsen, J. E. Ø., Otterå, O. H., Richter, K., Skagseth, Ø. A brief history of climate e the northern seas from the Last Glacial Maximum to global warming. *Quat. Sci. Rev*. 2014, *106*, 225–246.

Elverhoi A., Dowdeswell J. A., Funder S., Mangerud J., Stein R. (1998) Glacial and oceanic history of the Polar North Atlantic margins: an overview. *Quatern. Sci. Rev*. 1998, *17(1 – 3)*, 1 -10.

Fairbanks, R. G. A 17,000-years glacio-eustatic sea level record: influence of glacial melting rates on the Younger Dryas event and deep-ocean circulation. *Nature* 1989, *342*, 637–642.

Falardeau, J., de Vernal, A., Spielhagen, R. F., Paleoceanography of northeastern Fram Strait since the last glacial maximum: palynological evidence of large amplitudes changes. *Quat. Sci. Rev*., under review.

Forman S. L., Weihe R., Lubinski D., Tarasov G., Korsun S., Matishov G. (1997) Holocene relative sea-level history of Franz Josef Land, Russia. *Geol. Soc. Amer. Bull*. 1997, *109,* 1116-1133.

Forman S. L., Lubinski D. J., Zeeberg J. J., Polyak L., Miller G. H., Matishov G., Tarasov G. Observations of Postglacial emergence and Late Quaternary glaciation at selected localities on northern Novaya Zemlya, Russia. *Boreas* 1999, *28,* 133-145.

Fransner, O., Noormets, R., Chauhan, T., O'Regan, M. A., Jakobsson, M. Late Weichselian ice stream configuration, dynamics and retreat history in Albertini Trough, northern Svalbard margin (in prep.).

Gataullin, V., Mangerud, J., Svendsen, J. I. The extend of the Late Weichselian ice sheet in the southeastern Barents Sea. *Glob. Planet. Change* 2001, *31,* 453–474.

Hald M., Steinsund P. I. Benthic foraminifera and carbonate dissolution in the surface sediments of the Barents and Kara Seas. In: Stein R, Ivanov G, Levitan M, Fahl K (eds.) Surface sediment composition and sedimentary processes in the central Arctic Ocean and along the Eurasian Continental Margin. *Ber. Polarforsch* 1996, *212,* 285-307.

Hald, M., Kolstad, V., Polyak, L., Forman, S. L., Herlihy, F. A., Ivanov, G., Nesheretov, A. Late-glacial and Holocene paleoceanography and sedimentary environments in the St. Anna trough, Eurasian Arctic Ocean margin. *Paleoceanogr. Paleoclimatol. Palaecol.* 1999, *146,* 229–249.

Hebbeln, D, Dokken, T. M., Andersen, E. S., Hald, M., Elverhoi, A. Moisture supply for northern ice-sheet growth during the last glacial maximum. *Nature* 1994, *370,* 357-359.

Hillaire-Marcel, C., de Vernal, A. Stable isotope clue to episodic sea-ice formation in the glacial North Atlantic. *Earth Planet. Sci. Lett.* 2008, *268,* 143-150.

Hillaire-Marcel, C., Maccali, J., Not, C., Poirier, A. Geochemical and isotopic tracers of arctic sea ice sources and export with special attention to the Younger Dryas interval. *Quat. Sci. Rev.* 2013, *79,* 184-190.

Hogan, K. A., Dowdeswell, J. A., Noormets, R., Evans, J. Ó, Cofaigh, C., Jakobsson, M. Submarine landforms and ice-sheet flow in the Kvitøya Trough, northwestern Barents Sea. *Quat. Sci. Rev.* 2010a, *29,* 3545–3562.

Hogan, K. A., Dowdeswell, J. A., Noormets, R., Evans, J. Ó, Cofaigh, C. Geophysical and geological evidence for full-glacial ice flow and retreat of the Late Weichselian Ice Sheet from the waters around Kong Karls Land, eastern Svalbard. *Quat. Sci. Rev.* 2010b, *29,* 3563–3582.

Hughes, A. L. C., Gyllencreutz, R., Lohne, Ø. S., Mangerud, J., Svendsen, J. I. The last Eurasian ice sheets – a chronological database and time-slice reconstruction, DATED-1. *Boreas*. 2016, *45,* 1-45.

Husum, K., Hald, M. Modern foraminiferal distribution in the subarctic Malangen Fjord and adjoining shelf, Northern Norway. *J. Foram. Res*. 2004, *34,* 34–48.

Ivanova, E. V. Chapter 4. Influence of the Global Thermohaline Circulation on Paleoceanographic Events in the Eurasian Arctic Seas. In: *The Global Thermohaline Paleocirculation.* Springer: Dordrecht, Heidelberg, London, New York: 2009, pp 61-106.

Ivanova, E. V., Murdmaa, I. O., Duplessy, J. C., Paterne, M. Late Weichselian to Holocene Paleoenvironments in the Barents Sea. *Global Planet. Change*. 2002, *34 (3–4),* 69–78.

Ivanova, E. V., Murdmaa, I. O., Emelyanov, E. M., Seitkalieva, E. A., Radionova, E. P., Alekhina, G. N., Sloistov, S. M. Postglacial Paleoceanographic Environments in the Barents and Baltic Seas. *Oceanology*. 2016, *56 (1),* 118-130 (English Translation).

Ivanova, E., Murdmaa, I., de Vernal, A., Risebrobakken, B., Peyve, A., Bruce, C., Seitkalieva, E., Pisarev, S,. Postglacial paleoceanography of the northwestern Barents Sea: variations in the Atlantic Water inflow, iceberg calving and sea ice conditions. *Mar. Geol.,* under review.

Jones, G. A., Keigwin L. D. Evidence from Fram Strait (78° N) for early deglaciation. *Nature* 1988, *336*, 56-59.

Kaplin, P. A., Selivanov, A. O. *Modification of the Russian seas level and coasts development*. GEOS: Moscow, 1999, 289 p. (in Russian).

Klitgaard-Kristensen, D., Rasmussen, T. L., Koç, N. Palaeoceanographic changes in the northern Barents Sea during the last 16 000 years – new constraints on the last deglaciation of the Svalbard–Barents Sea Ice Sheet. *Boreas*. 2013, *42,* 798–813.

Knies J., Stein R. New aspects of organic carbon deposition and its paleoceanographic implications along the northern Barents Sea margin during the last 30,000 years. *Paleoceanography* 1998, *13(4),* 384-394.

Korsun, S., Hald M. Seasonal dynamics of benthic foraminifera in a glacially fed fjord of Svalbard, European Arctic. *J. Foramin. Res.* 2000, *30(4),* 251- 271.

Krapivner, R. B. Rapid subsidence of the Barents Sea shelf during the last 15-16 thousand years. *Geotectonics 2006*, *3*, 39-51 (English translation).

Krapivner R. B. Origin of weakly consolidated sediments from the Barents Sea shelf. *Lithology and mineral resouces* 2009, *1,* 96-100 (English translation).

Landvik, J. Y., Bondevik, S., Elverhøi, A., Fjeldskaar, W., Mangerud, J., Siegert, M. J., Salvigsen, O., Svendsen, J. I., Vorren, T. O. The last glacial maximum of Svalbard and the Barents Sea area: ice sheet extent and configuration. *Quat. Sci. Rev.* 1998, *17,* 43–75.

Lambeck K. Constrains on the late Weichselian ice sheet over the Barents Sea from observations of raised shorelines. *Quat. Sci. Rev*. 1995, *14,* 1-16.

Lubinski D. J., Korsun S., Polyak L., Forman S. L., Lehman S. J., Herlihy F. A., Miller G. H. The last deglaciation of the Franz Victoria Trough, northern Barents Sea. *Boreas* 1996, *25,* 89-100.

Lubinski, D. J., Polyak, L., Forman, S.L. Freshwater and Atlantic water inflows to the deep northern Barents and Kara seas since ca 13 $^{14}C$ ka: foraminifera and stable isotopes. *Quat. Sci. Rev.* 2001, *20,* 1851–1879.

McManus, J. F., Francois, R., Gherardi, J-M. et al. Collapse and rapid resumption of Atlantic meridional circulation linked to deglacial climate changes. *Nature* 2004, *428,* 834–837.

Murdmaa I. O., Ivanova E. V. Postglacial sedimentation history in the shelf depressions of the Barents Sea. *Lithology and Mineral Resources* 1999, *6,* 576-595 (English Translation).

Murdmaa I. O., Ivanova E. V., Duplessy J.-C., Levitan M. A., Khusid T. A., Alekseeva T. N., Bourtman M. V., Belousov M. A., Serova V. V. Facies System of the Central and Eastern Barents Sea since the Last Glaciation to Recent. *Mar. Geol*. 2006, *230*, 273-303.

Murton, J. B., Bateman, M. D., Dallimore, S. R., Teller, J. T., Yang, Z. Identification of Younger Dryas outburst flood path from Lake Agassiz to the Arctic Ocean. Nature 2010, *464,* 740-743.

Nørgaard-Pedersen N, Spielhagen R. F., Thiede J., Kassens H. Central Arctic surface ocean environment during the past 80,000 years. *Paleoceanography* 1998, *13(2),* 193-204.

Ottesen, D., Stokes, C. R., Risea, L., Olsen, L. Icesheet dynamics and ice streaming along the coastal parts of northern Norway, *Quat. Sci. Rev.* 2008, *27,* 922– 940.

Ottesen, D., Dowdeswell, J. A., Rise, L. Submarine landforms and the reconstruction of fast-flowing ice streams within a large Quaternary ice sheet: the 2500-km-long Norwegian-Svalbard margin (57°–80°N), *GSA Bull.* 2005, *117 (7/8),* 1033–1050.

Pavlidis, Yu. A., Polyakova, E. I. Late Pleistocene and Holocene depositional environments and paleoceanography of the Barents Sea: evidence from seismic and biostratigraphic data, *Mar. Geol.* 1997, *143,* 189–205.

Polyak, L., Mikhailov, V. Post-glacial environments of the southeastern Barents Sea: Foraminiferal evidence. In: Andrews JT, Austin WEN, Bergsten H, Jennings AE (eds) Late Quaternary paleoceanography of the North Atlantic margins. *Geological Society Special Publication,* 1996, 323-337.

Polyak, L., Solheim, A. Late- and postglacial environments in the northern Barents Sea, west of Franz Josef Land. *Polar Res.* 1994, *13,* 197-207.

Polyak, L., Lehman, S. J., Gataullin, V., Jull, A. J. T. Two-step deglaciation of the southeastern Barents Sea. *Geology* 1995, *23,* 567-571.

Polyak, L., Forman, S. L., Herlihy, F. A., Ivanov, G., Krinitsky, P. Late Weichselian deglacial history of the Svyataya (Saint) Anna Trough, northern Kara Sea, Arctic Russia. *Mar. Geol.* 1997, *143,* 169-188.

Polyak, L., Levitan, M., Gataullin, V., Khusid, T., Mikhailov, V., Mukhina, V. The impact of glaciation, river discharge, and sea-level change on Late Quaternary environments in the southwestern Kara Sea. Int. *J. Earth Sci.* 2000, 89, 550-562.

Poore P. Z., Osterman L., Curry W. B., Phillips R. L. Late Pleistocene and Holocene meltwater events in the western Arctic Ocean. *Geology* 1997, *27,* 759-762.

Rasmussen, T. L., Thomsen, E., Ślubowska, M. A., Jessen, S., Solheim, A., Koç, N. Paleoceanographic evolution of the SW Svalbard margin (76°N) since 20,000 14C yr BP. *Quat. Res.* 2007, *67,* 100-114.

Rasmussen, T. L., Forwick, M., Mackensen, A. Reconstruction of inflow of Atlantic Water to Isfjorden, Svalbard during the Holocene: Correlation to climate and seasonality. *Mar. Micropaleontol.* 2012, *94–95*, 80–90.

Rasmussen, T. L., Thomsen, E., Nielsen, T. Water mass exchange between the Nordic seas and the Arctic Ocean on millennial timescale during MIS 4–MIS 2. *Geochem., Geophys., Geosyst.* 2014, *15,* 530–544.

Risebrobakken, B., Moros, M., Ivanova, E., Chistyakova, N., Rosenberg, R. Climate and oceanographic variability in the SW Barents Sea during the Holocene. *The Holocene*. 2010, *20 (4),* 609–621.

Risebrobakken, B., Dokken, T., Smedsrud, L. H., Andersson, C., Jansen, E., Moros, M., Ivanova, E. V. Early Holocene temperature variability in the Nordic Seas: The role of oceanic heat advection versus changes in orbital forcing. *Paleoceanography* 2011, *26, PA4206.*

Sarnthein, M., Kreveld, van S., Erlenkauser, H., Grootes, P. M., Kucera, M., Pflaumann, U., Schulz, M. Centennial- to- millennial- scale periodicities of Holocene climate and sediment injections off the western Barents shelf. *Boreas* 2003, *32 (75), 447–461.*

Siegert, M. J., Dowdeswell, J. A. Late Weichselian iceberg, surface-melt and sediment production from the Eurasian Ice Sheet: results from numerical icesheet modeling, *Mar. Geol.* 2002, *188,* 109–127.

Ślubowska, M. A., Koç, N., Rasmussen, T. L., Klitgaard Kristensen, D. Changes in the flow of Atlantic Water into the Arctic Ocean since the last deglaciation: evidence from the northern Svalbard continental margin, 80°N. *Paleoceanography* 2005, *20, PA4014.*

Ślubowska-Woldengen, M., Rasmussen, T. L., Koç, N., Klitgaard-Kristensen, D., Nilsen, F., Solheim, A. Advection of Atlantic Water to the western and northern Svalbard shelf through the last 17.5 ka cal yr BP. *Quat. Sci. Rev*. 2007, *26,* 463–478.

Spielhagen R. F., Erlenkeuser H., Siegert C. History of freshwater runoff across the Laptev Sea (Arctic) during the last deglaciation. *Glob. Planet. Change* 2005, *48(1–3),* 187–207.

Stein R., Nam S. I., Schubert C., Vogt C., Futterer D., Heinemaeier J. The last deglasiation events in the eastern central Arctic Ocean. *Science* 1994, *264,* 692-696.

Steinsund P. I., Hald M. Recent calcium carbonate dissolution in the Barents Sea, Paleaoceanographic applications. *Mar. Geology* 1994, *117,* 303-316.

Svendsen, J. I., Alexanderson, H., Astakhov, V. I., Demidov, I., Dowdeswell, J. A., Funder, S., Gataullin, V., Henriksen, M., Hjort, C., Houmark-Nielsen, M., Hubberten, H. W., Ingólfsson, Ó., Jakobsson, M., Kjær, K. H., Larsen, E., Lokrantz, H., Lunkka, J. P., Lyså, A., Mangerud, J., Matioushkov, A., Murray, A., Möller, P., Niessen, F., Nikolskaya, O., Polyak, L., Saarnisto, M., Siegert, C., Siegert, M. J., Spielhagen, R. F., Stein, R. Late Quaternary ice-sheet history of northern Eurasia. *Quat. Sci. Rev*. 2004, *23,* 1229–1271.

Taldenkova, E., Bauch H. A., Stepanova, A., Ovsepyan, Y., Pogodina, I., Klyuvitkina, T., Nikolaev, S. Benthic and planktic community changes at the North Siberian margin in response to Atlantic water mass variability since last deglacial times. *Mar. Micropal.* 2012, *96–97,* 13–28.

Tarasov, L., Peltier, W. R. Arctic freshwater forcing of the Younger Dryas cold reversal. *Nature* 2005, *435,* 662-665.

Vorren T. O., Laberg J. S. Late glacial air temperature, oceanographic and ice sheet interactions in the southern Barents Sea region. In: Andrews JT, Austin WEN, Bergsten H, Jennings AE (eds) Late Quaternary paleoceanography of the North Atlantic margins. *Geo. Soc*. Spec. Publ. 1996, *111,* 275-287.

Werner, K., Spielhagen, R. F. Bauch, D., Christian Hass, H., Kandiano, E. Atlantic Water advection versus sea-ice advances in the eastern Fram Strait during the last 9 ka: Multiproxy evidence for a two-phase Holocene. *Paleoceanography* 2013, *29,* 283-295.

Wollenburg, J. E., Knies, J., Mackensen, A. High-resolution paleoproductivity fluctuations during the past 24 kyr as indicated by benthic foraminifera in the marginal Arctic Ocean. *Palaeogeogr., Palaeoclim., Palaeoecol.* 2004, *204,* 209–238.

Xiao X., Stein R., Fahl K. MIS 3 to MIS 1 temporal and LGM spatial variability in Arctic Ocean sea ice cover: Reconstruction from biomarkers. *Paleoceanography* 2015, *30,* 969–983.

Zeeberg J., Lubinski D. J., Forman S. L. Holocene Relative Sea-Level History of Novaya Zamlya, Russia, and Implications for Late Weichselian Ice-Sheet Loading. *Quat. Res*. 2001 *56(2),* 218-230.

**Reviewed By:**

*Yelena I. Polyakova*, Prof. Dr. Sc.
Geographical Faculty,
Lomonosov Moscow State University,
Moscow, Russia

# INDEX

## D



## E



## F



## G



## H



## I

## L



## M



## N



## O



## P



## Q

## R



## S



## T



## V



## W